# Distributions, Integral Transforms and Applications

# ANALYTICAL METHODS AND SPECIAL FUNCTIONS
## *An International Series of Monographs in Mathematics*

FOUNDING EDITOR: A.P. Prudnikov (Russia)
SERIES EDITORS: C.F. Dunkl (USA), H.-J. Glaeske (Germany) and M. Saigo (Japan)

This book is part of a series. The publisher will accept continuation orders which may be cancelled at any time and which provide for automatic billing and shipping of each title in the series upon publication. Please write for details.

# Distributions, Integral Transforms and Applications

Wladyslaw Kierat

and

Urszula Sztaba

CRC Press
Taylor & Francis Group
Boca Raton  London  New York

CRC Press is an imprint of the
Taylor & Francis Group, an **informa** business

A TAYLOR & FRANCIS BOOK

CRC Press
Taylor & Francis Group
6000 Broken Sound Parkway NW, Suite 300
Boca Raton, FL 33487-2742

First issued in paperback 2019

© 2003 by Taylor & Francis Group, LLC
CRC Press is an imprint of Taylor & Francis Group, an Informa business

No claim to original U.S. Government works

ISBN-13: 978-0-415-26958-2 (hbk)
ISBN-13: 978-0-367-39555-1 (pbk)

**Library of Congress Cataloging-in-Publication Data**

Catalog record is available from the Library of Congress

**Visit the Taylor & Francis Web site at
http://www.taylorandfrancis.com**

**and the CRC Press Web site at
http://www.crcpress.com**

# CONTENTS

# PREFACE

This book has been written as an approachable introduction to the theory of distributions and integral transforms. The principal intention of our book is to emphasize the remarkable connections between distributional theory, classical analysis and theory of differential equations. We use the theory of the Lebesgue integral as a fundamental tool in the proofs of many theorems. The theory is developed from its beginning to the point where many of the fundamental and deep theorems are proved, for example the Schwartz kernel theorem and the Malgrange–Ehrenpreis theorem. We particularly want to show some practical means of using the theory of distributions in the cases (for example Fourier analysis), when methods of classical analysis are insufficient. This book naturally covers general theory, examples and applications. We also try to answer the main questions related to the topics presented in the text. The material presented in our book will be understandable to a student who has completed a course of advanced calculus. The book has been made self-contained by inclusion of the necessary facts from functional analysis in the Appendix. There are elementary approaches to the theory of distributions considering distributions as limits of sequences of functions. The theoretical foundations in these approaches are usually presented only in a simplified manner insufficient for applications. In this volume we give Schwartz's description of distributions as linear continuous forms on topological vector spaces. This presentation of distributions is flexible in applications in various branches of mathematics.

This book is divided into seven chapters.

In Chapter 1 we define distributions and introduce some fundamental notions as regular distributions, weak derivatives of locally integrable functions, derivatives of distributions and the convergence of distributions.

Chapter 2 treats the question of local properties of distributions.

Chapter 3 is devoted to tensor products and convolution products of distributions.

In Chapter 4 we give some applications of the theory of distributions to differential equations.

In the next two chapters we deal with integral transforms such as the Cauchy transforms, Fourier transforms and Hilbert transforms.

In the last chapter we present some aspects of the theory of orthogonal expansions of some classes of distributions.

The basic definitions and theorems of measure theory are taken from Rudin's book: Real and Complex Analysis ([23]). Theorems, lemmas, definitions and formulas are numbered separately in each section, for example Theorem 2 in Section 10 of Chapter 1 will be denoted by Theorem 1.10.2.

*Wladyslaw Kierat and Urszula Sztaba*

# CHAPTER 1

# DEFINITIONS AND PRELIMINARIES

## 1.1. The spaces $\mathcal{D}$ and $\mathcal{D}'$

We will use the following symbols:

$\mathbb{N}^n$ — set of all non-negative integer points of $\mathbb{R}^n$,

$e_i = (0, \ldots, 0, 1, 0, \ldots, 0)$, , where number 1 is on $i$-th position,

$\alpha \leq \beta$, $\alpha$, $\beta \in \mathbb{N}^n$ means $\alpha_i \leq \beta_i$ for $i = 1, \ldots, n$,

$\alpha < \beta$ means $\alpha \leq \beta$ and $\alpha_i < \beta_i$ for some $i = 1, \ldots, n$,

$|\alpha| = \alpha_1 + \alpha_2 \cdots + \alpha_n$,

$\alpha! = \alpha_1! \alpha_2! \cdots \alpha_n!$,

$\binom{\alpha}{\beta} = \frac{\alpha!}{\beta! \, (\alpha - \beta)!}$,

$\partial^\alpha = \frac{\partial^{|\alpha|}}{\partial_{x_1}^{\alpha_1} \ldots \partial_{x_n}^{\alpha_n}} = \frac{\partial^{|\alpha|}}{\partial^\alpha x}$,

$\operatorname{supp} \varphi = \overline{\{x : \varphi(x) \neq 0\}}$ .

Let $\Omega$ be an open set in $\mathbb{R}^n$. We will denote by $\mathcal{D}(\Omega)$ the set of all infinitely differentiable functions with compact supports contained in $\Omega$. These functions will be called smooth. If $\Omega = \mathbb{R}^n$ we will write $\mathcal{D}$ instead of $\mathcal{D}(\mathbb{R}^n)$. (In another cases, when functions constituting a functional space are defined on whole $\mathcal{R}^n$, then we will proceed analogously.) Obviously, $\mathcal{D}(\Omega)$ is a vector space.

EXAMPLE 1.1.1. The following function

$$\varphi(x) = \begin{cases} \exp(|x|^2 - 1)^{-1} & \text{for } |x| < 1, \ |x| = (x_1^2 + \cdots + x_n^2)^{\frac{1}{2}}, \\ 0 & \text{for } |x| \geq 1 \end{cases}$$

belongs to $\mathcal{D}$.

We introduce a convergence of sequences in $\mathcal{D}(\Omega)$.

DEFINITION 1.1.1. We say that a sequence $(\varphi_\nu)$, $\varphi_\nu \in \mathcal{D}(\Omega)$ converges to $\varphi \in \mathcal{D}(\Omega)$ iff

   (i) there exists a compact set $K \subset \Omega$ such that $\operatorname{supp} \varphi_\nu \subset K$ for each $\nu \in \mathbb{N}$ and
   (ii) the sequence $\partial^\alpha \varphi_\nu$ converges to $\partial^\alpha \varphi$ uniformly on $K$ for each $\alpha \in \mathbb{N}^n$.

Of course, the sequence $(\varphi_\nu)$ converges to $\varphi$ in $\mathcal{D}(\Omega)$ iff the sequence $(\varphi_\nu - \varphi)$ converges to zero in $\mathcal{D}(\Omega)$.

DEFINITION 1.1.2. A linear form defined on $\mathcal{D}(\Omega)$ continuous with respect to the above convergence is said to be a distribution.

The set of distributions will be denoted by $\mathcal{D}'(\Omega)$.

THEOREM 1.1.1. *A linear form* $\Lambda$ *defined on* $\mathcal{D}(\Omega)$ *is continuous iff for every compact set* $K \subset \Omega$ *there exist a constant* $C > 0$ *and an index* $m$ *such that*

$$|\Lambda(\varphi)| \leq C\|\varphi\|_{m,K}, \qquad\qquad (1.1.1)$$

*where* $\|\varphi\|_{m,K} = \max_{|\alpha| \leq m} \sup_{x \in K} \left|\frac{\partial^{|\alpha|}}{\partial x^\alpha} \varphi(x)\right|$ *and* supp $\varphi \subset K$.

PROOF. The sufficiency is evident. We shall now prove the necessity. Assume that $\Lambda \in \mathcal{D}'(\Omega)$ and (1.1.1) does not hold for some compact set $K \subset \Omega$. Therefore for each $\nu \in \mathbb{N}$ there exists a function $\varphi_\nu \in \mathcal{D}(\Omega)$ such that supp $\varphi_\nu \subset K$ and $|\Lambda(\varphi_\nu)| > \nu\|\varphi_\nu\|_{\nu,K}$. Obviously, one can choose $\varphi_\nu$ such that $\Lambda(\varphi_\nu) = 1$ for each $\nu \in \mathbb{N}$. From this we obtain

$$1 = \Lambda(\varphi_\nu) > \nu\|\varphi_\nu\|_{\nu,K},$$

consequently $\|\varphi_\nu\|_{\nu,K} < \frac{1}{\nu}$. Hence

$$\|\varphi_{\nu+\mu}\|_{\mu,K} \leq \|\varphi_{\nu+\mu}\|_{\nu+\mu,K} \leq \frac{1}{\nu+\mu} \leq \frac{1}{\nu}.$$

From this it follows that $\|\varphi_{\nu+\mu}\|_{\mu,K} \to 0$ as $\nu \to \infty$ for fixed $\mu$. Therefore $\varphi_\nu$ converges to zero as $\nu \to \infty$ in $\mathcal{D}(\Omega)$. This contradicts the continuity of $\Lambda$. Thus, the proof is achieved. $\square$

EXAMPLE 1.1.2. Let $f \in L^1_{loc}$. We take

$$\Lambda_f(\varphi) := \int\limits_{\mathbb{R}^n} f\varphi.$$

Of course, $|\Lambda_f(\varphi)| \leq \|f\|_{L^1(K)}\|\varphi\|_{0,K}$ if supp $\varphi \subset K$. Therefore $\Lambda_f$ is in $\mathcal{D}'$. The distribution $\Lambda_f$ is said to be regular.

EXAMPLE 1.1.3. Put $\delta(\varphi) := \varphi(0)$ for $\varphi \in \mathcal{D}$. It is very easily seen that (1.1.1) holds. We shall now show that $\delta$ is not a regular distribution. Indeed, let $\varphi$ be as in Example 1.1.1. Put $\varphi_\nu(x) := \varphi(\nu x)$. Obviously, $\varphi_\nu(0) = e^{-1}$. Moreover, for each $x \neq 0$, $\lim_{\nu \to \infty} \varphi_\nu(x) = 0$. Assume that $\delta(\varphi) = \int_{\mathbb{R}^n} f\varphi$ for some $f \in L^1_{loc}$ and $\varphi \in \mathcal{D}$. Obviously, $e^{-1} = \delta(\varphi_\nu) = \int_{\mathbb{R}^n} f\varphi_\nu$. Note that $\int_{\mathbb{R}^n} f\varphi_\nu$ tends to zero as $\nu \to \infty$. This contradicts our assumption. Therefore $\delta$ is not a regular distribution. The distribution $\delta$ is called Dirac measure concentrated in the zero point.

EXAMPLE 1.1.4. Let $\varphi$ be in $\mathcal{D}$ and supp $\varphi \subset [-a, a]$. Put

$$\Lambda(\varphi) = \frac{1}{(\cdot)}(\varphi) := \lim_{\epsilon \to 0} \int\limits_{|x| > \epsilon} \frac{\varphi(x)}{x}\, dx.$$

Let us take a new function

$$\psi(x) = \begin{cases} \frac{\varphi(x) - \varphi(0)}{x} & \text{for } x \neq 0, \\ \varphi'(0) & \text{for } x = 0. \end{cases}$$

Clearly, $\psi$ is continuous and $\varphi(x) = \varphi(0) + x\psi(x)$ for $x \in \mathbb{R}$. Of course, $\psi(x) = \varphi'(\Theta_x \cdot x)$, $0 < \Theta_x < 1$ for $x \neq 0$. This implies that

$$|\psi(x)| \leq \max_{x \in [-a,a]} |\varphi'(x)| \leq \|\varphi\|_{1,[-a,a]}$$

for $x \in \mathbb{R}$. Note that

$$\int_{|x|>\epsilon} \frac{\varphi(x)}{x} \, dx = \int_{\epsilon<|x|\leq a} \frac{\varphi(x)}{x} \, dx = \int_{\epsilon<|x|\leq a} \psi(x) \, dx.$$

By the Lebesgue dominated convergence theorem we have

$$\Lambda(\varphi) = \lim_{\epsilon \to 0} \int_{|x|>\epsilon} \frac{\varphi(x)}{x} \, dx = \int_{|x|<a} \psi(x) \, dx.$$

Hence we obtain the following inequality

$$|\Lambda(\varphi)| \leq 2a\|\varphi\|_{1,[-a,a]}$$

if $\operatorname{supp} \varphi \subset [-a, a]$. Thus, we have shown that $\Lambda$ is a distribution. The distribution $\frac{1}{(\cdot)}$ is called the Cauchy finite part of the integral $\int_{\mathbb{R}} \frac{1}{(\cdot)}$.

## 1.2. Approximation lemmas

Suppose that $\psi \in L^1$ and $\int_{\mathbb{R}^n} \psi = 1$. Let $\psi_\epsilon(x) = \epsilon^{-n} \psi\left(\frac{x}{\epsilon}\right)$, $\epsilon > 0$ and $f \in L^p$, $1 \leq p \leq \infty$. Put

$$(f * \psi_\epsilon)(y) := \int_{\mathbb{R}^n} f(x-y)\psi_\epsilon(x) \, dx.$$

We prove the following

LEMMA 1.2.1. *If $f \in L^p$, $1 \leq p < \infty$, then $f * \psi_\epsilon \to f$ as $\epsilon \to 0$ in $L^p$, if $f \in L^\infty$ and $f$ is uniformly continuous on an open set $V \subset \mathbb{R}^n$, then $f * \psi_\epsilon$ almost uniformly converges to $f$ as $\epsilon \to 0$ on $V$.*

PROOF. We now recall Minkowski's inequality

$$\left[\int_G \left(\int_\Omega |F(x,y)| \, dy\right)^p dx\right]^{\frac{1}{p}} \leq \int_\Omega \left(\int_G |F(x,y)|^p \, dx\right)^{\frac{1}{p}} dy, \qquad (1.2.1)$$

where $\Omega \subset \mathbb{R}^n$, $G \subset \mathbb{R}^m$ and $1 \leq p < \infty$. It is easy to check that $\int_{\mathbb{R}^n} \psi_\epsilon(x) \, dx = 1$ for all $\epsilon > 0$. Hence

$$(f * \psi_\epsilon)(x) - f(x) = \int_{\mathbb{R}^n} (f(x-y) - f(x)) \, \psi_\epsilon(y) \, dy$$

$$= \int_{\mathbb{R}^n} (f(x-\epsilon y) - f(x)) \, \psi(y) \, dy.$$

Because of (1.2.1) we have

$$\left( \int\limits_{\mathbb{R}^n} \left| \int\limits_{\mathbb{R}^n} (f(x - \epsilon y) - f(x))\psi(y)\, dy \right|^p dx \right)^{\frac{1}{p}}$$

$$\leq \left( \int\limits_{\mathbb{R}^n} \left( \int\limits_{\mathbb{R}^n} |(f(x - \epsilon y) - f(x))\psi(y)|\, dy \right)^p dx \right)^{\frac{1}{p}}$$

$$\leq \int\limits_{\mathbb{R}^n} \left( \int\limits_{\mathbb{R}^n} |f(x - \epsilon y) - f(x)|^p |\psi(y)|^p dx \right)^{\frac{1}{p}} dy$$

$$= \int\limits_{\mathbb{R}^n} |\psi(y)| \left( \int\limits_{\mathbb{R}^n} |f(x - \epsilon y) - f(x)|^p\, dx \right)^{\frac{1}{p}} dy$$

$$= \int\limits_{\mathbb{R}^n} |\psi(y)|\, \|f(\cdot - \epsilon y) - f(\cdot)\|_{L^p}\, dy.$$

Note that $\|f(\cdot - \epsilon y) - f(\cdot)\|_{L^p} \leq 2\|f\|_{L^p}$ and $\|f(\cdot - \epsilon y) - f(\cdot)\|_{L^p} \to 0$ as $\epsilon \to 0$ for each $y \in \mathbb{R}^n$. Hence, by the Lebesgue dominated convergence theorem we infer that $f * \psi_\epsilon \to f$ as $\epsilon \to 0$ in $L^p$. Thus, the proof of the first part of our theorem is finished. On the other hand suppose that $f \in L^\infty$ and $f$ is uniformly continuous on an open set $V$. Having given $\delta > 0$, choose a compact set $W \subset \mathbb{R}^n$ so that $\int_{\mathbb{R}^n - W} |\psi| < \delta$. Then we have

$$\sup_{x \in V_0} |(f * \psi_\epsilon)(x) - f(x)|$$

$$\leq \sup_{x \in V_0} \int\limits_{W} |f(x - \epsilon y) - f(x)||\psi(y)|\, dy + \sup_{x \in V_0} \int\limits_{\mathbb{R}^n - W} |f(x - \epsilon y) - f(x)|\, |\psi(y)|\, dy$$

$$\leq \eta \int\limits_{\mathbb{R}^n} |\psi(y)|\, dy + 2\delta\|f\|_{L^\infty}$$

for $V_0 \Subset V$, where this symbol means that the closure $\overline{V_0}$ of $V_0$ is compact and $\overline{V_0} \subset V$. Note that $\eta$ is small for small $\epsilon$. This finishes the proof. $\qquad\square$

REMARK 1.2.1. In the preceding proof, the appeal to the Lebesgue dominated convergence theorem may seem to be illegitimate since the Lebesgue dominated convergence theorem deals only with countable sequences of functions. However, it does enable us to conclude that

$$\int\limits_{\mathbb{R}^n} |\psi(y)|\, \|f(\cdot - \epsilon_\nu y) - f(\cdot)\|_{L^p}\, dy \to 0$$

as $\epsilon_\nu \to 0$ for every sequence $(\epsilon_\nu)$ which converges to zero and it says exactly that

$$\int\limits_{\mathbb{R}^n} |\psi(y)|\, \|f(\cdot - \epsilon y) - f(\cdot)\|_{L^p}\, dy \to 0$$

as $\epsilon \to 0$. We shall encounter similar situations again and shall apply convergence theorems to them without further comment.

COROLLARY 1.2.1. *If* $\psi \in L^1$, $\int_{\mathbb{R}^n} \psi = 1$ *and* $\psi_\epsilon(x) = \epsilon^{-n}\psi\left(\frac{x}{\epsilon}\right)$, *then* $\int_{\mathbb{R}^n} \psi_\epsilon \varphi \to \varphi(0)$ *as* $\epsilon \to 0$ *for* $\varphi \in \mathcal{D}$.

PROOF. Put $\check{\varphi}(x) := \varphi(-x)$. Then

$$\int_{\mathbb{R}^n} \psi_\epsilon \varphi = \int_{\mathbb{R}^n} \check{\varphi}(-x)\psi_\epsilon(x)\,dx = (\check{\varphi} * \psi_\epsilon)(0) \to \check{\varphi}(0) = \varphi(0)$$

as $\epsilon \to 0$. This statement finishes the proof. $\qquad\square$

A very similar lemma to Lemma 1.2.1 is true for arbitrary open sets in $\mathbb{R}^n$. We shall now introduce some notations. Let

$$B_r(\Omega) = \{x : \operatorname{dist}(x, \Omega) < r\},$$

in particular, $B_r(0) = B_r(\{0\}) = \{x : |x| < r\}$. Let $\psi$ be in $L^1$, $\operatorname{supp}\psi \subset B_1(0)$, $\int_{\mathbb{R}^n} \psi = 1$ and $f \in L^p(\Omega)$. Put by definition

$$f_\epsilon(x) := \epsilon^{-n} \int_\Omega \psi\left(\frac{x-y}{\epsilon}\right) f(y)\,dy.$$

We shall now show that $f_\epsilon \in L^p(\Omega)$ and $\operatorname{supp} f_\epsilon \subset \overline{B_r(\Omega)}$ for $\epsilon < r$. Indeed, let us extend $f$ on $\mathbb{R}^n$ putting $\tilde{f}(x) := f(x)$ for $x \in \Omega$ and $\tilde{f}(x) = 0$ for $x \notin \Omega$. Take by definition $\psi_\epsilon(x) = \epsilon^{-n}\psi\left(\frac{x}{\epsilon}\right)$. Then we have $f_\epsilon = \psi_\epsilon * \tilde{f}$. By the Young inequality (see (5.5.1)) we infer that $f_\epsilon \in L^p(\mathbb{R}^n)$. Therefore $f_\epsilon \in L^p(\Omega)$. It is easy to verify that $f_\epsilon(x) = 0$ for $x \notin \overline{B_r(\Omega)}$. By virtue of Lemma 1.2.1 we get

$$\|f_\epsilon - \tilde{f}\|_{L^p(\mathbb{R}^n)} \to 0$$

as $\epsilon \to 0$. From this, we infer that $\|f_\epsilon - f\|_{L^p(\Omega)} \to 0$ as $\epsilon \to 0$.

Our consideration leads to the following

LEMMA 1.2.2. *If* $f \in L^p(\Omega)$, $\psi \in L^1$, $\operatorname{supp}\psi \subset B_1(0)$, $\int \psi = 1$ *and*

$$f_\epsilon(x) = \epsilon^{-n} \int_\Omega \psi\left(\frac{x-y}{\epsilon}\right) f(y)\,dy$$

*then* $f_\epsilon \in L^p(\Omega)$, $f_\epsilon$ *tends to* $f$ *as* $\epsilon \to 0$ *in* $L^p(\Omega)$ *and* $\operatorname{supp} f_\epsilon \subset \overline{B_r(\Omega)}$ *for* $\epsilon < r$.

## 1.3. Regularizations of functions

Let $f$ be in $L^1_{loc}$ and $\psi$ be in $\mathcal{D}$.

DEFINITION 1.3.1. The function $f_\psi = \psi * f$ is said to be the regularization of $f$ by $\psi$.

THEOREM 1.3.1. *If* $f \in L^1_{loc}$ *and* $\psi \in \mathcal{D}$ *then* $f_\psi$ *is in* $C^\infty$ *and* $\partial^\alpha f_\psi = \partial^\alpha \psi * f$ *for* $\alpha \in \mathbb{N}^n$. *Moreover, if* $f \in L^p$, $1 \le p < \infty$, $\int_{\mathbb{R}^n} \psi(x)\,dx = 1$ *and* $\psi_\epsilon(x) = \epsilon^{-n}\psi\left(\frac{x}{\epsilon}\right)$ *then* $f_{\psi_\epsilon}$ *tends to* $f$ *as* $\epsilon \to 0$ *in* $L^p$.

PROOF. Since $(f * \psi)(x) = \int_{\mathbb{R}^n} \psi(x - y) f(y)\, dy$ therefore

$$f_\psi(x + \xi) - f_\psi(x) = \int_{\mathbb{R}^n} (\psi(x + \xi - y) - \psi(x - y)) f(y)\, dy.$$

Note that the supports of $\psi(x+\xi-\cdot) - \psi(x-\cdot)$ are contained in a compact set when $x$ is fixed and $|\xi| < \delta$. Since $f$ is in $L^1_{loc}$, we conclude by the Lebesgue dominated convergence theorem that $f_\psi(x + \xi) - f_\psi(x)$ tends to zero as $\xi \to 0$. Therefore $f_\psi$ is continuous. Of course,

$$(f_\psi(x + he_i) - f_\psi(x))h^{-1} = \int_{\mathbb{R}^n} \frac{\partial}{\partial x_i} \psi(x + \Theta h e_i - y) f(y)\, dy,$$

where $e_i = (0, \ldots 0, 1, 0, \ldots 0)$. Repeating the previous consideration we infer that $(f_\psi(x + he_i) - f_\psi(x))h^{-1}$ tends to $\left( \frac{\partial}{\partial x_i} \psi * f \right)(x)$ as $h \to 0$. Now we proceed by induction. The second part of the lemma is a consequence of Lemma 1.2.1. □

DEFINITION 1.3.2. We say that the function $\varphi$ with a compact support belongs to $C_c(\Omega)$ if it is continuous and supp $\varphi \subset \Omega$.

THEOREM 1.3.2. *Let $\psi \in \mathcal{D}$, supp $\psi \subset B_1(0)$ and $\int_{\mathbb{R}^n} \psi = 1$. If $f$ is in $C_c(\Omega)$, then there exists $\epsilon_0 > 0$ such that $f_{\psi_\epsilon} \in \mathcal{D}(\Omega)$ for $\epsilon < \epsilon_0$ and $f_{\psi_\epsilon}$ tends to $f$ uniformly in $\Omega$.*

PROOF. Without loss of generality we assume that $f \in C_c(\mathbb{R}^n)$. Namely, one can take $f(x) = 0$ for $x \notin \text{supp } f$. Let $x \notin \overline{B_\epsilon(\text{supp } f)}$, then $|x - y| > \epsilon$ for each $y$ in supp $f$ and $\psi_\epsilon(x - y) = 0$. Hence we have $(\psi_\epsilon * f)(x) = 0$. This implies that

$$\text{supp } f_{\psi_\epsilon} \subset \overline{B_\epsilon(\text{supp } f)}.$$

Therefore if $\epsilon_0 < \text{dist}(\text{supp } f, \partial\Omega)$ then $f_{\psi_\epsilon} \in \mathcal{D}(\Omega)$ for $\epsilon < \epsilon_0$. In view of Lemma 1.2.1, $f_{\psi_\epsilon}$ converges uniformly to $f$ in $\Omega$. This finishes the proof. □

### 1.4. Du Bois-Rymond's lemma

We know that every locally integrable function $f$ determines a regular distribution $\Lambda_f$ by equality $\Lambda(\varphi) = \int f\varphi$ (see Section 1.1). We shall now show that the mapping $f \to \Lambda_f$ is an injection from $L^1_{loc}(\Omega)$ into $\mathcal{D}'(\Omega)$. This fact will be an immediate consequence of the following

LEMMA 1.4.1 (du Bois-Rymond). *If $f$ is in $L^1_{loc}(\Omega)$ and $\int_\Omega f\varphi = 0$ for $\varphi \in \mathcal{D}(\Omega)$ then $f(x) = 0$ almost everywhere in $\Omega$.*

PROOF. Let $K \Subset \Omega$. Put

$$\text{sgn } f(x) := \begin{cases} \frac{f(x)}{|f(x)|} & \text{if } f(x) \neq 0 \\ 0 & \text{if } f(x) = 0. \end{cases}$$

Note that $|f(x)| = \frac{f(x)\overline{f(x)}}{|f(x)|} = f(x) \text{sgn } \overline{f(x)}$. Let $K \Subset \Omega$. Denote by $\chi_K$ the characteristic function of $K$. Further, put $g(x) := \chi_K \text{sgn } \overline{f(x)}$. Obviously, $|g(x)| \leq$

1 for $x \in \Omega$. Because of Lusin's lemma (see [23]) for arbitrary $\delta > 0$ there exists $\psi \in C_c(\Omega)$ such that $|\psi(x)| \leq 1$ for $x \in \Omega$ and $\mu(A) < \delta$, where

$$A = \{x : x \in \Omega, \ \psi(x) \neq g(x)\}.$$

Of course, the following equalities

$$\int\limits_K |f(x)| dx = \int\limits_\Omega \chi_K f(x) \operatorname{sgn} \overline{f(x)} \, dx$$

$$= \int\limits_\Omega f(x)\psi(x) \, dx + \int\limits_\Omega f(x)(\chi_K \operatorname{sgn} \overline{f(x)} - \psi(x)) \, dx$$

hold. Assuming that $\int_\Omega f\varphi = 0$ for $\varphi \in \mathcal{D}(\Omega)$ and taking into account Theorem 1.3.2, we infer that $\int_\Omega f\psi = 0$ for $\psi \in C_c(\Omega)$. From this it follows that

$$\int\limits_K |f(x)| \, dx = \int\limits_A f(x)(\chi_K \operatorname{sgn} \overline{f(x)} - \psi(x)) \, dx,$$

where $\mu(A) < \delta$. Since $K \Subset \Omega$ and $f \in L^1_{loc}(\Omega)$ therefore for every $\epsilon > 0$ one can choose $\delta > 0$ so that $\int_A f(x)(\chi_K \operatorname{sgn} \overline{f(x)} - \psi(x)) \, dx < \epsilon$ if $\mu(A) < \delta$. This statement finishes the proof. □

## 1.5. Some density theorem

THEOREM 1.5.1. $\mathcal{D}(\Omega)$ is dense in $L^p(\Omega)$ if $1 \leq p < \infty$.

PROOF. Because of Theorem 1.3.2 we only need to show that $C_c(\Omega)$ is dense in $L^p(\Omega)$. Let $f$ be in $L^p(\Omega)$ and $\epsilon > 0$. We shall show that there exists a function $\varphi \in C_c(\Omega)$ such that $\|f - \varphi\|_{L^p(\Omega)} < \epsilon$. Setting $f = f_1 - f_2 + i(f_3 - f_4)$ where each $f_j$, $j = 1, 2, 3, 4$ is real valued and nonnegative, we can restrict ourselves without loss of generality to proving a reduced form of the theorem. Namely, we assume that $f$ is in $L^p(\Omega)$ and it takes nonnegative values. Choose a monotonically increasing sequence $(s_\nu)$ of nonnegative simple functions converging pointwise to $f$ in $\Omega$. Since

$$0 \leq s_\nu(x) \leq f(x)$$

therefore $s_\nu \in L^p(\Omega)$ and

$$(f(x) - s_\nu(x))^p \leq f^p(x).$$

From this, by the Lebesgue dominated convergence theorem it follows that $s_\nu$ tends to $f$ in $L^p(\Omega)$. Hence we have $\|s_\nu - f\|_{L^p} < \frac{\epsilon}{2}$ for $\nu \geq \nu_0$. Since $s_\nu$ is a simple function therefore the supp of $s_\nu$ must have finite volume. By Lusin's lemma there exists $\varphi \in C_c(\Omega)$ such that

$$|\varphi(x)| \leq \|s_{\nu_0}\|_{L^\infty}$$

and

$$\operatorname{vol}(\{x : x \in \Omega, \ s_{\nu_0}(x) \neq \varphi(x)\}) < \left(\frac{\epsilon}{4} \left(\|s_{\nu_0}\|_{L^\infty}\right)^{-1}\right)^p.$$

Hence we obtain

$$\|s_{\nu_0} - \varphi\|_{L^p} \leq \|s_{\nu_0} - \varphi\|_{L^\infty} (\text{vol}\{x : x \in \Omega, \ s_{\nu_0}(x) \neq \varphi(x)\})^{\frac{1}{p}}$$

$$\leq 2\|s_{\nu_0}\|_{L^\infty} \frac{\epsilon}{4} (\|s_{\nu_0}\|_{L^\infty})^{-1} = \frac{\epsilon}{2}$$

Finally we have $\|f - \varphi\|_{L^p} < \epsilon$. Thus, the theorem is proved. $\square$

## 1.6. Distributional convergence

Let $\Lambda_\alpha$, $\Lambda$ be in $\mathcal{D}'(\Omega)$, $\alpha \in \mathbb{R}$.

DEFINITION 1.6.1. We say that $\Lambda_\alpha$ distributionally converges to $\Lambda$ as $\alpha \to \alpha_0$ if $\lim_{\alpha \to \alpha_0} \Lambda_\alpha(\varphi) = \Lambda(\varphi)$ for each $\varphi \in \mathcal{D}(\Omega)$.

In particular, if $\nu \in \mathbb{N}$ then we take $\lim_{\nu \to \infty} \Lambda_\nu = \lim_{\frac{1}{\nu} \to 0} \Lambda_\nu = \Lambda$ if $\lim_{\nu \to \infty} \Lambda_\nu(\varphi) = \Lambda(\varphi)$ for each $\varphi \in \mathcal{D}(\Omega)$.

EXAMPLE 1.6.1. The following sequences

$$\frac{1}{2}\nu e^{-\nu|x|} \qquad \text{(Picard)}, \tag{1.6.1}$$

$$\frac{\nu}{e^{\nu x} + e^{-\nu x}} \qquad \text{(Stieltjes)}, \tag{1.6.2}$$

$$\frac{1}{\pi} \frac{\nu}{1 + (\nu x)^2} \qquad \text{(Cauchy)}, \tag{1.6.3}$$

$$\frac{\nu}{\sqrt{2\pi}} e^{-\frac{1}{2}(\nu x)^2} \qquad \text{(Gauss)} \tag{1.6.4}$$

distributionally converge to $\delta$. See Corollary 1.2.1.

We shall now show that for each $\varphi$ in $\mathcal{D}$

$$\lim_{\epsilon \to 0+} \int_{\mathbb{R}} \frac{\varphi(x)}{x + i\epsilon} \, dx = -i\pi\delta(\varphi) + \frac{1}{(\cdot)}(\varphi) \tag{1.6.5}$$

and

$$\lim_{\epsilon \to 0+} \int_{\mathbb{R}} \frac{\varphi(x)}{x - i\epsilon} \, dx = i\pi\delta(\varphi) + \frac{1}{(\cdot)}(\varphi), \tag{1.6.6}$$

where $\frac{1}{(\cdot)}(\varphi) = \lim_{\epsilon \to 0} \int_{|x| \geq \epsilon} \frac{\varphi(x)}{x} \, dx$ (see Example 1.1.4). The equalities (1.6.5) and (1.6.6) are said to be the Sochocki–Plemelj formulas.

Let $\varphi$ be in $\mathcal{D}$ and supp $\varphi \subset [-a, a]$. We know that $\varphi$ may be written as follows $\varphi(x) = \varphi(0) + x\psi(x)$, where $\psi$ is a continuous function. Moreover, $\frac{1}{(\cdot)}(\varphi) = \int_{-a}^{a} \psi$ (see Example 1.1.4). Hence we have

$$\int_{\mathbb{R}} \frac{\varphi(x)}{x + i\epsilon} \, dx = \int_{-a}^{a} \frac{\varphi(0)}{x + i\epsilon} \, dx + \int_{-a}^{a} \frac{x\psi(x)}{x + i\epsilon} \, dx.$$

Note that

$$\int_{-a}^{a} \frac{\varphi(0)}{x+i\epsilon}\, dx = \varphi(0) \int_{-a}^{a} \frac{x}{x^2+\epsilon^2}\, dx - \pi i \varphi(0)\frac{1}{\pi}\int_{-a}^{a}\frac{\epsilon}{x^2+\epsilon^2}dx$$

$$= -\pi i \varphi(0)\frac{1}{\pi}\int_{\frac{-a}{\epsilon}}^{\frac{a}{\epsilon}} \frac{1}{t^2+1}\, dt.$$

Finally we have

$$\lim_{\epsilon\to 0^+}\int_{\mathbb{R}} \frac{\varphi(0)}{x+i\epsilon}\, dx = -\pi i \varphi(0) = -\pi i \delta(\varphi).$$

Of course,

$$\lim_{\epsilon\to 0}\int_{-a}^{a} \frac{x\psi(x)}{x+i\epsilon}\, dx = \int_{-a}^{a} \psi(x)\, dx.$$

This statement finishes the proof of (1.6.5). Analogously one can prove (1.6.6).

EXAMPLE 1.6.2. An easy computation shows that $\int_{\mathbb{R}} 2x^2(1+x^2)^{-2}\, dx = \pi$. Take $\psi(x) = 2\pi^{-1}x^2(1+x^2)^{-2}$ and $\psi_\nu(x) = \nu\psi(\nu x)$. The sequence $(\psi_\nu)$ converges to $\delta$ (see Corollary 1.2.1). On the other hand $\psi_\nu(x)$ tends to zero as $\nu \to \infty$ for each $x \in \mathbb{R}$. This example indicates that the pointwise convergence is not the distributional convergence in general.

EXAMPLE 1.6.3. Let $f_\nu$, $f$ be in $L^1_{loc}(\Omega)$ and the sequence $(f_\nu)$ converges to $f$ as $\nu \to \infty$ in the sense of $L^1_{loc}(\Omega)$. It is easily seen that $\int_\Omega f_\nu \varphi$ converges to $\int_\Omega f\varphi$ as $\nu \to \infty$ for each $\varphi \in \mathcal{D}(\Omega)$. This means that the sequence $(f_\nu)$ distributionally converges to $f$.

## 1.7. Algebraic operations on distributions

Let $S$ and $T$ be in $\mathcal{D}'(\Omega)$.

DEFINITION 1.7.1. The expression $S + T$ defined in the following way

$$(S+T)(\varphi) := S(\varphi) + T(\varphi), \quad \varphi \in \mathcal{D}(\Omega)$$

is called the sum of the distributions $S$ and $T$.

Obviously, $S+T$ is also a continuous linear form on $\mathcal{D}(\Omega)$. Moreover, if $S_f$ and $T_g$ are regular distributions corresponding to the functions $f$ and $g$, then $S_f + T_g$ is also a regular distribution and it corresponds to the function $f + g$.

There is not any natural way to define the product of two arbitrary distributions. Nevertheless, it is possible to define the product of any distribution $\Lambda$ by an infinitely differentiable function $g$. Note that the product of an infinitely differentiable function $g$ and a function $\varphi$ from $\mathcal{D}(\Omega)$ belongs to $\mathcal{D}(\Omega)$. Moreover, if the sequence $(\varphi_\nu)$, $\varphi_\nu \in \mathcal{D}(\Omega)$ converges to zero in $\mathcal{D}(\Omega)$ then the sequence $(g\varphi_\nu)$ also converges to zero in $\mathcal{D}(\Omega)$.

Let $\Lambda$ be a distribution. Then the mapping $\varphi \to \Lambda(g\varphi)$ defines a new distribution on $\mathcal{D}(\Omega)$.

DEFINITION 1.7.2. The distribution $\varphi \to \Lambda(g\varphi)$ is said to be the product of the distribution $\Lambda$ and the function $g$.

For the regular distribution $\Lambda_f$ corresponding to the locally integrable function $f$, multiplication by $g$ corresponds to multiplication of $f$ by $g$ in the usual sense.

## 1.8. Linear transformations in the space of the independent variables

Let $A$ be a linear nonsingular transformation in $\mathbb{R}^n$ and $A^{-1}$ be its inverse. Assuming $f$ to be locally integrable and letting $\varphi$ be a function in $\mathcal{D}$, by changing variables in the integral $\int_{\mathbb{R}^n} f(A^{-1}x)\varphi(x)\,dx$ we obtain

$$\int_{\mathbb{R}^n} f(A^{-1}x)\varphi(x)\,dx = |A| \int_{\mathbb{R}^n} f(y)\varphi(Ay)\,dy, \qquad (1.8.1)$$

where $y = A^{-1}x$ and $|A|$ is the absolute value of the determinant of the matrix of the transformation $A$. Equality (1.8.1) can be written as follows

$$\Lambda_{f\circ A^{-1}}(\varphi) = |A|\Lambda_f(\varphi \circ A), \qquad (1.8.2)$$

where $\Lambda_{f\circ A^{-1}}$ and $\Lambda_f$ are the regular distributions corresponding to the locally integrable functions $f \circ A^{-1}$ and $f$, respectively. Taking into account (1.8.2) one can define the linear transformation $A^{-1}$ for any distribution $\Lambda \in \mathcal{D}'$ in the following way

$$A^{-1}\Lambda(\varphi) := |A|\Lambda(\varphi \circ A), \qquad (1.8.3)$$

where $|A|$ is the absolute value of the determinant of the matrix of the transformation $A$. If $\Lambda_f$ is the regular distribution corresponding to the function $f$ from $L^1_{loc}$ then $A^{-1}\Lambda_f = \Lambda_{f\circ A^{-1}}$.

We now describe some examples of very useful linear transformations.

EXAMPLE 1.8.1 (Reflection in the origin). Taking $x = Ay =: -y$, $\check{\varphi}(y) = \varphi(-y)$ we have $\check{\Lambda}(\varphi) = \Lambda(\check{\varphi})$. If $\Lambda_f$ is the regular distribution corresponding to $f \in L^1_{loc}$ we get $\check{\Lambda}_f(\varphi) = \int_{\mathbb{R}^n} f\check{\varphi} = \int_{\mathbb{R}^n} \check{f}\varphi$. Therefore $\check{\Lambda}_f = \Lambda_{\check{f}}$.

EXAMPLE 1.8.2 (Similarity transformation). Put $x = Ay := \alpha y$, $\alpha > 0$. According to (1.8.3) similarity transform $S_{\frac{1}{\alpha}}$ is defined by

$$S_{\frac{1}{\alpha}}\Lambda(\varphi) := \alpha^n \Lambda(\varphi(\alpha\,\cdot)).$$

If $\Lambda_f$ is the regular distribution corresponding to the function $f$ belonging to $L^1_{loc}$ then

$$S_{\frac{1}{\alpha}}\Lambda_f(\varphi) = \alpha^n \int_{\mathbb{R}^n} f(y)\varphi(\alpha y)\,dy = \int_{\mathbb{R}^n} f\left(\frac{x}{\alpha}\right)\varphi(x)\,dx.$$

Finally we have $S_{\frac{1}{\alpha}}\Lambda_f = \Lambda_{f(\frac{\cdot}{\alpha})}$.

EXAMPLE 1.8.3 (Orthogonal transformation). Let $A$ be an orthogonal transformation in $\mathbb{R}^n$. Further, let $\Lambda$ be in $\mathcal{D}'$ then taking into account (1.8.3) we have

$$A^{-1}\Lambda(\varphi) = \Lambda(\varphi \circ A)$$

If $\Lambda_f$ is the regular distribution corresponding to the function $f$ from $L^1_{loc}$ then we have

$$A^{-1}\Lambda_f = \Lambda_{f \circ A^{-1}} = \Lambda_{f \circ A^T},$$

where $A^T$ is the transpose matrix of $A$.

Although translation through the vector $h$ of independent variables is not a linear transformation in $\mathbb{R}^n$, we shall define the translation operator $\tau_h$ in the same way as the previous examples.

EXAMPLE 1.8.4 (Translation through the vector $h$). The translation operator $\tau_{-h}$ we define by (1.8.3) putting

$$\tau_{-h}\Lambda(\varphi) := \Lambda(\tau_h(\varphi)) = \Lambda(\varphi(\cdot - h)).$$

One can verify that if $\Lambda_f$ is the regular distribution corresponding to $f \in L^1_{loc}$, then $\tau_{-h}\Lambda_f = \Lambda_{\tau_{-h}f}$.

## 1.9. Differentiation of distributions

Let $\varphi$ and $f$ be in $\mathcal{D}(\Omega)$ and $C^{(m)}(\Omega)$ respectively, $\Omega \subset \mathbb{R}^n$. Integrating by parts and recalling that $\varphi$ is in $\mathcal{D}(\Omega)$ we arrive at

$$\int_\Omega \frac{\partial}{\partial x_i} f(x)\varphi(x)\,dx = -\int_\Omega f(x)\frac{\partial}{\partial x_i}\varphi(x)\,dx.$$

By induction we have

$$\int_\Omega \frac{\partial^{|\alpha|}}{\partial x^\alpha} f(x)\varphi(x)\,dx = (-1)^{|\alpha|}\int_\Omega f(x)\frac{\partial^{|\alpha|}}{\partial x^\alpha}\varphi(x)\,dx \qquad (1.9.1)$$

for $|\alpha| \le m$. Of course, the mapping

$$\varphi \to (-1)^{|\alpha|}\int_\Omega f(x)\frac{\partial^{|\alpha|}}{\partial x^\alpha}\varphi(x)\,dx$$

is a linear continuous form on $\mathcal{D}(\Omega)$. From (1.9.1) it follows that the above mapping is the regular distribution corresponding to the function $\partial^\alpha f$, where $\partial^\alpha f$ denotes the $\alpha$-partial derivative of $f$. We shall now use equation (1.9.1) to define the partial derivatives of a distribution.

DEFINITION 1.9.1. Let $\varphi$ be in $\mathcal{D}(\Omega)$, then the linear form defined by

$$\varphi \to (-1)^{|\alpha|}\Lambda(\partial^\alpha \varphi)$$

will be called the $\alpha$-derivative of $\Lambda$. This form will be denoted by $D^\alpha\Lambda$.

From now on, for simplicity of notations we write $f$ instead of $\Lambda_f$ if $\Lambda_f$ is the regular distribution corresponding to the function $f$. Note that if $f$ is in $C^{(m)}(\Omega)$, then $D^\alpha f = \partial^\alpha f$ for $|\alpha| \le m$. Moreover, for each distribution $\Lambda \in \mathcal{D}'(\Omega)$, $D^\alpha\Lambda = D^\beta\Lambda$, where the operators $D^\alpha$ and $D^\beta$ have only the different order of differentiation.

EXAMPLE 1.9.1. Let $v$ be in $L^1_{loc}(\mathbb{R})$ and put $u(x,y) = v(x+y)$. We shall now show that the function $u$ fulfils the wave equation

$$D^{(2,0)}u - D^{(0,2)}u = 0 \qquad (1.9.2)$$

in $\mathbb{R}^2$. We have to show that

$$\iint_{\mathbb{R}^2} v(x+y) \left( \frac{\partial^2}{\partial x^2} \varphi(x,y) - \frac{\partial^2}{\partial y^2} \varphi(x,y) \right) dx\, dy = 0 \qquad (1.9.3)$$

for $\varphi \in \mathcal{D}(\mathbb{R}^2)$. In order to do it we change the variables in the above integral putting $\xi = x - y$, $\eta = x + y$. Therefore

$$x = \frac{\xi + \eta}{2}, \qquad y = \frac{\eta - \xi}{2}.$$

An easy computation shows that

$$\frac{\partial^2}{\partial \xi \partial \eta} \varphi \left( \frac{\xi + \eta}{2}, \frac{\eta - \xi}{2} \right) = \frac{1}{4} \left( \frac{\partial^2}{\partial x^2} \varphi(x,y) - \frac{\partial^2}{\partial y^2} \varphi(x,y) \right).$$

After changing variables in the integral (1.9.3) we have

$$\iint_{\mathbb{R}^2} v(x+y) \left( \frac{\partial^2}{\partial x^2} \varphi(x,y) - \frac{\partial^2}{\partial y^2} \varphi(x,y) \right) dx\, dy$$

$$= 2 \iint_{\mathbb{R}^2} v(\eta) \frac{\partial^2}{\partial \eta \partial \xi} \varphi \left( \frac{\xi + \eta}{2}, \frac{\eta - \xi}{2} \right) d\xi\, d\eta$$

$$= 2 \int_{\mathbb{R}} v(\eta) \frac{\partial}{\partial \eta} \left( \int_{\mathbb{R}} \frac{\partial}{\partial \xi} \varphi \left( \frac{\xi + \eta}{2}, \frac{\eta - \xi}{2} \right) d\xi \right) d\eta.$$

Note that the function $\varphi \left( \frac{\cdot + \eta}{2}, \frac{\eta - \cdot}{2} \right)$ vanishes without a bounded interval depending on $\eta$. This implies that the interior integral is equal to zero for each $\eta$. Therefore (1.9.3) holds.

We shall now show that differentiation of distributions is a linear continuous operation in $\mathcal{D}'(\Omega)$.

THEOREM 1.9.1. If $\Lambda_\nu$ and $\Lambda$ are in $\mathcal{D}'(\Omega)$ and $\Lambda_\nu$ tends to $\Lambda$ as $\nu \to \infty$ in $\mathcal{D}'(\Omega)$ then $D^\alpha \Lambda_\nu$ also tends to $D^\alpha \Lambda$ as $\nu \to \infty$.

PROOF. The theorem is an immediate consequence of the definition of the derivatives of distributions. $\qquad \square$

## 1.10. Weak derivatives of locally integrable functions

DEFINITION 1.10.1. Let $f$ be in $L^1_{loc}(\Omega)$. If there exists a function $v$ belonging to $L^1_{loc}(\Omega)$ such that

$$\int_\Omega v(x)\varphi(x)\, dx = (-1)^\alpha \int_\Omega f(x) \frac{\partial^{|\alpha|}}{\partial x^\alpha} \varphi(x)\, dx$$

for $\varphi \in \mathcal{D}(\Omega)$, then we say that $v$ is $\alpha$-weak derivative of $f$.

In this case we shall write $\partial^\alpha f = v$ instead of $D^\alpha f = v$. That means that the distributional derivative $D^\alpha f$ of the locally integrable function $f$ is the regular distribution corresponding to $v$. In some situations, when $\alpha \in \mathbb{N}$, the weak or distributional derivatives will be denoted traditionally.

THEOREM 1.10.1. *If the locally integrable functions $v_1$ and $v_2$ are $\alpha$-weak derivatives of $f \in L^1_{loc}(\Omega)$ then $v_1(x) = v_2(x)$ almost everywhere on $\Omega$ (we write $v_1 = v_2$ a.e.).*

PROOF. The theorem is an immediate consequence of du Bois-Rymond's lemma. □

EXAMPLE 1.10.1. Let $H(x) = 0$ if $x < 0$ and $H(x) = 1$ if $x \geq 0$ (Heaviside step function). Therefore $D^1 H(\varphi) = -\int_{-\infty}^{\infty} H(x)\varphi'(x)\,dx = \varphi(0) = \delta(\varphi)$. The distributional derivative $D^1 H$ of $H$ is not a weak derivative of $H$.

EXAMPLE 1.10.2. Put $f(x,y) = 0$ if $x \geq y$ and $f(x,y) = y - x$ if $x < y$ for $(x,y) \in \mathbb{R}^2$. The function $f$ has the classical derivatives outside the set $A = \{(x,y) : x = y,\ x, y \in \mathbb{R}\}$. Namely

$$\frac{\partial}{\partial x} f(x,y) = 0 \quad \text{if} \quad x > y \qquad \text{and} \qquad \frac{\partial}{\partial x} f(x,y) = -1 \quad \text{if} \quad x < y.$$

Similarly

$$\frac{\partial}{\partial y} f(x,y) = 0 \quad \text{if} \quad x > y \qquad \text{and} \qquad \frac{\partial}{\partial y} f(x,y) = 1 \quad \text{if} \quad x < y.$$

We shall now show that $D^{(1,0)} f = \partial^{(1,0)} f$. Indeed, note that

$$-\int_{-\infty}^{\infty}\int_{-\infty}^{\infty} f(x,y)\frac{\partial}{\partial x}\varphi(x,y)\,dx\,dy = -\int_{-\infty}^{\infty}\left(\int_{-\infty}^{y}(y-x)\frac{\partial}{\partial x}\varphi(x,y)\,dx\right)dy$$

$$= -\int_{-\infty}^{\infty}\left(\int_{-\infty}^{y}\varphi(x,y)\,dx\right)dy = \int_{-\infty}^{\infty}\int_{-\infty}^{\infty}\frac{\partial}{\partial x}f(x,y)\varphi(x,y)\,dx\,dy.$$

Analogously one can show that $D^{(0,1)} f = \partial^{(0,1)} f$.

EXAMPLE 1.10.3. Let $f$ be absolutely continuous on $[\alpha, \beta]$. Then $f(x) = f(\alpha) + \int_\alpha^x f'(t)\,dt$ and $f' \in L^1(\alpha, \beta)$. Note that

$$-\int_\alpha^\beta f(t)\varphi'(t)\,dt = \int_\alpha^\beta f'(t)\varphi(t)\,dt$$

if $\varphi$ is in $\mathcal{D}(\alpha, \beta)$. This implies that $f'$ is the weak derivative of $f$.

Further, we shall need the following

LEMMA 1.10.1. *Let $f$ be a continuous function defined on the interval $(\alpha, \beta)$. If $\int_\alpha^\beta f(t)\varphi'(t)\,dt = 0$ for every $\varphi \in \mathcal{D}(\alpha, \beta)$ then $f$ is a constant function.*

PROOF. In the first place we show that the function $\theta \in \mathcal{D}(\alpha, \beta)$ is the derivative of some function $\psi$ belonging to $\mathcal{D}(\alpha, \beta)$ iff $\int_\alpha^\beta \theta(t)\,dt = 0$.

Necessity is evident. Assume that $\theta \in \mathcal{D}(\alpha, \beta)$ and $\int_\alpha^\beta \theta(t)\,dt = 0$. Put $\psi(x) := \int_\alpha^x \theta(t)\,dt$. We shall now show that $\operatorname{supp}\psi \subset (\alpha + \epsilon, \beta - \epsilon)$ for some $\epsilon > 0$. Indeed, let $\operatorname{supp}\theta \subset (\alpha + \epsilon, \beta - \epsilon)$. First of all $\psi(x) = 0$ if $x \in (\alpha, \alpha + \epsilon]$. For $x \in [\beta - \epsilon, \beta)$ we have $\psi(x) = \int_{\alpha+\epsilon}^{\beta-\epsilon} \theta(t)\,dt + \int_{\beta-\epsilon}^x \theta(t)\,dt$. This implies that $\int_\alpha^x \theta(t)\,dt = 0$ for $x \in [\beta - \epsilon, \beta)$. Therefore $\operatorname{supp}\psi \subset (\alpha + \epsilon, \beta - \epsilon)$. Of course, $\psi' = \theta$.

Further, each function $\varphi$ in $\mathcal{D}(\alpha, \beta)$ can be represented as follows

$$\varphi(x) = \varphi_0(x) \int_\alpha^\beta \varphi(t)\,dt + \theta(x),$$

where $\varphi_0, \theta \in \mathcal{D}(\alpha, \beta)$ and $\int_\alpha^\beta \varphi_0 = 1$, $\int_\alpha^\beta \theta = 0$. In fact, one can put

$$\theta(x) := \varphi(x) - \varphi_0(x) \int_\alpha^\beta \varphi,$$

where $\int_\alpha^\beta \varphi_0 = 1$. We see at once that $\int_\alpha^\beta \theta = 0$. In accordance with the previous consideration, there exists a function $\psi$ in $\mathcal{D}(\alpha, \beta)$ such that

$$\varphi(x) = \varphi_0(x) \int_\alpha^\beta \varphi + \psi'(x).$$

Assume that $f$ is a continuous function, then $\int_\alpha^\beta f\varphi = \int_\alpha^\beta \varphi \int_\alpha^\beta f\varphi_0 + \int_\alpha^\beta \psi' f$. From the assumption, we have $\int_\alpha^\beta \psi' f = 0$. Therefore $\int_\alpha^\beta f\varphi = c \int_\alpha^\beta \varphi$, where $c = \int_\alpha^\beta f\varphi_0$. Because of the continuity of $f$ in view of du Bois-Rymond's lemma it follows that $f(x) = c$ for $x \in (\alpha, \beta)$. $\qquad\square$

We are now in a position to prove the following

THEOREM 1.10.2. *If $f$ is a continuous function on $(\alpha, \beta)$ but it is not absolutely continuous, then it has no weak derivative.*

PROOF. In fact, assume that the locally integrable function $v$ is the weak derivative of $f$, then for each function $\varphi$ in $\mathcal{D}(\alpha, \beta)$ we have

$$-\int_\alpha^\beta f\varphi' = \int_\alpha^\beta v\varphi.$$

Put $V(x) = \int_\alpha^x v(t)\,dt$, then $-\int_\alpha^\beta v\varphi = \int_\alpha^\beta V\varphi'$. Therefore $\int_\alpha^\beta (f - V)\varphi' = 0$. By virtue of Lemma 1.10.1, $f = V + c$. This contradicts our assumption. $\qquad\square$

REMARK 1.10.1. Let $\varphi$ and $\Lambda$ be in $\mathcal{D}(\alpha, \beta)$ and $\mathcal{D}'(\alpha, \beta)$ respectively, $(\alpha, \beta) \subset \mathbb{R}$. Assume that the derivative $D\Lambda$ of $\Lambda$ is equal to zero, this means $\Lambda(\varphi') = 0$ for each $\varphi \in \mathcal{D}(\alpha, \beta)$. We know that for each $\varphi$ in $\mathcal{D}(\alpha, \beta)$ there exist two functions $\varphi_0$ and $\theta$ in $\mathcal{D}(\alpha, \beta)$ such that

$$\varphi = \varphi_0 \int\limits_\alpha^\beta \varphi + \theta \quad \text{and} \quad \int\limits_\alpha^\beta \varphi_0 = 1, \quad \int\limits_\alpha^\beta \theta = 0.$$

The function $\varphi_0$ can be chosen independently on $\varphi$. Hence we have

$$\Lambda(\varphi) = \Lambda(\varphi_0) \int\limits_\alpha^\beta \varphi + \Lambda(\theta).$$

Let $\psi \in \mathcal{D}(\alpha, \beta)$ and $\psi' = \theta$. Since $\Lambda(\psi') = 0$ therefore $\Lambda(\varphi) = \int_\alpha^\beta \Lambda(\varphi_0)\varphi$. From this, it follows that $\Lambda$ is a regular distribution generated by the constant function $\Lambda(\varphi_0)$. This distribution fulfils the differential equation $Dx = 0$. Therefore each distributional solution of the differential equation $Dx = 0$ is a regular distribution generated by a constant function. Note that it satisfies this equation at each point $x$ of $(\alpha, \beta)$. One can replace the constant function $\Lambda(\varphi_0)$ by another function, for example putting $f(x) := 0$ if $x$ is a rational number, but for $x$ which is not rational it takes $f(x) := \Lambda(\varphi_0)$. Of course,

$$\int\limits_\alpha^\beta f\varphi' = \int\limits_\alpha^\beta \Lambda(\varphi_0)\varphi'.$$

From this, it follows that the function $f$ is a distributional solution of differential equation $Dx = 0$. Obviously, the function $f$ is not differentiable in any point of $(\alpha, \beta)$. The above observation indicates that there exists an essential difference between two statements: the function $f$ satisfies a differential equation everywhere or almost everywhere on a set $\Omega$ and it fulfils this differential equation in the distributional sense on $\Omega$.

EXAMPLE 1.10.4. We shall now show that $D \ln |\cdot| = \frac{1}{(\cdot)}$. Indeed, let $\varphi$ be in $\mathcal{D}$. Then we have

$$D \ln |\cdot| (\varphi) = - \int\limits_{-\infty}^\infty \ln |x| \varphi'(x) \, dx = - \int\limits_{-\infty}^0 \ln(-x)\varphi'(x) \, dx - \int\limits_0^\infty \ln x \, \varphi'(x) \, dx.$$

Of course,

$$\int\limits_0^\infty \ln x \, \varphi'(x) \, dx = \lim_{\epsilon \to 0^+} \int\limits_\epsilon^\infty \ln x \, \varphi'(x) \, dx.$$

Note that

$$- \int\limits_{-\infty}^{-\epsilon} \ln(-x)\varphi'(x) \, dx = -\varphi(-\epsilon) \ln \epsilon + \int\limits_{-\infty}^{-\epsilon} \frac{\varphi(x)}{x} \, dx$$

and

$$-\int_{\epsilon}^{\infty} \ln x \, \varphi'(x) \, dx = \varphi(\epsilon) \ln \epsilon + \int_{\epsilon}^{\infty} \frac{\varphi(x)}{x} \, dx.$$

It is easily seen that $\lim_{\epsilon \to 0+} (\varphi(\epsilon) - \varphi(-\epsilon)) \ln \epsilon = 0$. From this it follows that

$$-\int_{-\infty}^{\infty} \ln|x| \varphi'(x) \, dx = \lim_{\epsilon \to 0+} \int_{|x| \geq \epsilon} \frac{\varphi(x)}{x} \, dx.$$

Finally we have $D \ln |\cdot| = \frac{1}{(\cdot)}$.

## 1.11. Local Sobolev spaces

In this section there will be introduced local Sobolev spaces as a natural generalization of the spaces $L_{loc}^p(\Omega)$. Assume that $f$ has the weak derivatives $\partial^\alpha f$ up to order $m$ on $\Omega$. We shall now define a family of seminorms in this way

$$|f\|_{m,p,K} = \left( \sum_{|\alpha| \leq m} \|\partial^\alpha f\|_{L^p(K)}^p \right)^{\frac{1}{p}} \quad \text{if} \quad 1 \leq p < \infty, \tag{1.11.1}$$

$$\|f\|_{m,\infty,K} = \max_{|\alpha| \leq m} \|\partial^\alpha f\|_{L^\infty(K)} \quad \text{if} \quad p = \infty, \tag{1.11.2}$$

where $K$ is a compact set contained in $\Omega$. Let

$$W_{loc}^{m,p}(\Omega) := \{f : \partial^\alpha f \in L_{loc}^p(\Omega) \text{ for } |\alpha| \leq m, \, \partial^\alpha f \text{ is the weak derivative of } f\}.$$

Of course, $W_{loc}^{m,p}(\Omega)$ is a vector space.

DEFINITION 1.11.1. Vector spaces $W_{loc}^{m,p}(\Omega)$ equipped with the appropriate seminorms (1.11.1) or (1.11.2), where $K$ goes through the family of all compact sets contained in $\Omega$ are called local Sobolev spaces over $\Omega$.

Obviously, $W_{loc}^{0,p}(\Omega) = L_{loc}^p(\Omega)$. The purpose of our considerations is to make a description of functions belonging to $W_{loc}^{m,p}(\Omega)$, when $1 \leq p < \infty$. Let $\varphi$ be in $\mathcal{D}$. If $f$ is in $L_{loc}^p(\Omega)$, then the following integral

$$\int_{\Omega} \varphi(x - y) f(y) \, dy$$

is meaningful for each $x \in \mathbb{R}^n$. The function $f_\varphi$, defined by

$$f_\varphi(x) := \int_{\Omega} \varphi(x - y) f(y) \, dy$$

is called the regularization of $f$ by means of $\varphi$. Taking $\tilde{f}(x) = f(x)$ if $x \in \Omega$ and $\tilde{f}(x) = 0$ if $x \notin \Omega$ we obtain

$$f_\varphi(x) = (\varphi * \tilde{f})(x) = \int_{\mathbb{R}^n} \varphi(x - y) \tilde{f}(y) \, dy.$$

By virtue of Theorem 1.3.1 we infer that $f_\varphi$ is smooth and

$$\frac{\partial^{|\alpha|}}{\partial x^\alpha} f_\varphi(x) = \int_\Omega \frac{\partial^{|\alpha|}}{\partial x^\alpha} \varphi(x - y) f(y)\, dy. \tag{1.11.3}$$

Let $\psi \geq 0$ be in $\mathcal{D}$, supp $\psi \subset B_1(0)$ and $\int \psi = 1$. Put $\psi_\epsilon(x) = \epsilon^{-n} \psi\left(\frac{x}{\epsilon}\right)$, $\epsilon > 0$. In the sequel, we shall need the regularization $f_\epsilon$ given by

$$f_\epsilon(x) := f_{\psi_\epsilon}(x) = \int_\Omega \psi_\epsilon(x - y) f(y)\, dy = \int_{\mathbb{R}^n} \psi_\epsilon(x - y) \tilde{f}(y)\, dy. \tag{1.11.4}$$

At the beginning, we present two lemmas.

LEMMA 1.11.1. *If $f$ is in $L^1_{loc}(\Omega)$ and $\partial^\alpha f$ is its weak derivative, then for the regularizations $f_\epsilon$ and $(\partial^\alpha f)_\epsilon$ given by (1.11.4) we have*

$$\frac{\partial^{|\alpha|}}{\partial x^\alpha} f_\epsilon(x) = (\partial^\alpha f)_\epsilon(x), \tag{1.11.5}$$

*if $x \in \Omega_\epsilon = \{x : x \in \Omega,\ \mathrm{dist}(x, \partial\Omega) > \epsilon\}$.*

PROOF. In accordance with the definition of weak derivative we get

$$(\partial^\alpha f)_\epsilon(x) = \int_\Omega \psi_\epsilon(x - y) \partial^\alpha f(y)\, dy$$

$$= (-1)^{|\alpha|} \int_\Omega \frac{\partial^{|\alpha|}}{\partial y^\alpha} \psi_\epsilon(x - y) f(y)\, dy$$

$$= \int_\Omega \frac{\partial^{|\alpha|}}{\partial x^\alpha} \psi_\epsilon(x - y) f(y)\, dy.$$

By (1.11.3) we get (1.11.5) for $x \in \Omega_\epsilon$. Thus, the proof is finished. $\square$

LEMMA 1.11.2. *If $f$ is in $L^p_{loc}(\Omega)$ and $1 \leq p < \infty$, then $f_\epsilon$, given by (1.11.4) tends to $f$ as $\epsilon \to 0$ in $L^p_{loc}(\Omega)$.*

PROOF. After a change of variables in the integral $\int_{\mathbb{R}^n} \psi_\epsilon(x - y) \tilde{f}(y)\, dy$ we obtain

$$f_\epsilon(x) = \int_{B_1(0)} \psi(z) \tilde{f}(x - \epsilon z)\, dz.$$

Let $K \subset \Omega$ be a compact set. Using Minkowski's inequality we get the following estimation

$$\left( \int\limits_K |f_\epsilon(x)|^p dx \right)^{\frac{1}{p}} \leq \left( \int\limits_K \left( \int\limits_{B_1(0)} \psi(z) |\tilde{f}(x - \epsilon z)| dz \right)^p dx \right)^{\frac{1}{p}}$$

$$\leq \int\limits_{B_1(0)} \psi(z) \left( \int\limits_K |\tilde{f}(x - \epsilon z)|^p dx \right)^{\frac{1}{p}} dz$$

$$\leq \|\tilde{f}\|_{L^p(B_\epsilon(K))}.$$

Assume that $\epsilon < d = \text{dist}(K, \partial\Omega)$. Then we get

$$\|f_\epsilon\|_{L^p(K)} \leq \|f\|_{L^p(B_\epsilon(K))}. \tag{1.11.6}$$

Without loss of generality one can assume that $K = \overline{\Omega'}$, $\Omega' \Subset \Omega$. Because of the density of $C_c(\Omega')$ in $L^p(\Omega')$, for fixed $\eta > 0$ there exists a function $g$ in $C_c(\Omega')$ such that $\|f - g\|_{L^p(K)} < \frac{\eta}{4}$. By virtue of Theorem 1.3.2, we infer that $\|g - g_\epsilon\|_{L^p(K)} < \frac{\eta}{4}$ for small $\epsilon$, where $g_\epsilon$ is defined by (1.11.4). Note that $f_\epsilon - g_\epsilon = (f - g)_\epsilon$. By (1.11.6) we have

$$\|f_\epsilon - g_\epsilon\|_{L^p(K)} \leq \|f - g\|_{L^p(B_\epsilon(K))} \leq \|f - g\|_{L^p(K)} + \|f\|_{L^p(B_\epsilon(K)-K)}.$$

Since $f \in L^p_{loc}(\Omega)$, therefore

$$\|f\|_{L^p(B_\epsilon(K)-K)} < \frac{\eta}{4}$$

for small $\epsilon$. Now, one can make the following estimation

$$\|f - f_\epsilon\|_{L^p(K)} \leq \|f - g\|_{L^p(K)} + \|g - g_\epsilon\|_{L^p(K)} + \|g_\epsilon - f_\epsilon\|_{L^p(K)} < \eta$$

for small $\epsilon$. Thus, the proof of the lemma is finished. $\qquad\square$

THEOREM 1.11.1. *Let $f$ be in $L^p_{loc}(\Omega)$ and let $D^\alpha f$ be its distributional derivative. $D^\alpha f$ is weak derivative of $f$ and belongs to $L^p_{loc}(\Omega)$ iff there exists a sequence $(f_\nu)$, $f_\nu \in C^\infty(\Omega)$ convergent to $f$ in $L^p_{loc}(\Omega)$ such that the sequence $(\partial^\alpha f_\nu)$ is a Cauchy sequence in $L^p_{loc}(\Omega)$.*

PROOF OF SUFFICIENCY. Assume that the sequence $(f_\nu)$, $f_\nu \in C^\infty(\Omega)$ converges to $f$ in $L^p_{loc}(\Omega)$ and that the sequence $(\partial^\alpha f_\nu)$ is a Cauchy sequence in $L^p_{loc}(\Omega)$. By continuity of differentiation, $\partial^\alpha f_\nu$ tends to $D^\alpha f$ as $\nu \to \infty$ in $\mathcal{D}'(\Omega)$. Since $(\partial^\alpha f_\nu)$ is a Cauchy sequence in $L^p_{loc}(\Omega)$, therefore there exists a function $v_\alpha \in L^p_{loc}(\Omega)$ such that

$$\lim_{\nu \to \infty} \int\limits_\Omega \partial^\alpha f_\nu \varphi = \int\limits_\Omega v_\alpha \varphi = D^\alpha f(\varphi)$$

for $\varphi \in \mathcal{D}(\Omega)$. Finally we get

$$D^\alpha f(\varphi) = (-1)^{|\alpha|} \int\limits_\Omega f \partial^\alpha \varphi = \int\limits_\Omega v_\alpha \varphi.$$

This means that $D^\alpha f = v_\alpha$ is the weak derivative of $f$.                    □

PROOF OF NECESSITY. Let $K \subset \Omega$ be a compact set and let $d = \operatorname{dist}(K, \partial\Omega)$. Assume that $\partial^\alpha f$ is the weak derivative of $f$ in $\Omega$. Take the sequence $\left(f_{\frac{1}{\nu}}\right)$, $\nu \in \mathbb{N}$, $f_{\frac{1}{\nu}}$ defined by (1.11.4). Then by Lemma 1.11.1 we have

$$\frac{\partial^{|\alpha|}}{\partial x^\alpha} f_{\frac{1}{\nu}}(x) = (\partial^\alpha f)_{\frac{1}{\nu}}(x)$$

for $x \in K$, if $\frac{1}{\nu} < d$. From this and Lemma 1.11.2, it follows that $f_{\frac{1}{\nu}}$ and $(\partial^\alpha f)_{\frac{1}{\nu}}$ tend to $f$ and $\partial^\alpha f$ as $\nu \to \infty$, respectively under the norm $\|\cdot\|_{L^p(K)}$. This statement finishes the proof of necessity.                    □

From our consideration the following corollaries follow.

COROLLARY 1.11.1. $C^\infty(\Omega)$ is dense in $W_{loc}^{m,p}(\Omega)$ if $1 \le p < \infty$.

COROLLARY 1.11.2. The space $W_{loc}^{m,p}(\Omega)$, $1 \le p < \infty$ is complete in the following sense: if $(f_\nu)$, $f_\nu \in W_{loc}^{m,p}(\Omega)$ is a Cauchy sequence in $W_{loc}^{m,p}(\Omega)$, then there exists a function $f$ in $W_{loc}^{m,p}(\Omega)$ such that $\lim_{\nu\to\infty} f_\nu = f$ in the sense of $W_{loc}^{m,p}(\Omega)$.

PROOF. Compare with Theorem 1.12.1.                    □

COROLLARY 1.11.3. If $f$ is in $W_{loc}^{m,p}(\Omega)$, then
  (i) $D^\beta(\partial^\alpha f) = \partial^{(\alpha+\beta)} f$ if $|\alpha + \beta| \le m$,
  (ii) $\partial^\alpha f = \partial^\beta f$ if $|\alpha| \le m$, where $\partial^\alpha$ and $\partial^\beta$ have only the different order of differentiation.

## 1.12. Sobolev spaces

We shall now introduce Sobolev spaces of integer order and establish some of their properties. Similarly as in the previous section we denote by $W^{m,p}(\Omega)$ the set $\{f : f \in L^p(\Omega),\ 1 \le p \le \infty,\ \partial^\alpha f \in L^p(\Omega)$ for $|\alpha| \le m$, where $\partial^\alpha f$ is the weak derivative of $f\}$. Obviously, $W^{m,p}(\Omega)$ is a vector space.

DEFINITION 1.12.1. The vector spaces $W^{m,p}(\Omega)$ equipped with the norms

$$\|f\|_{m,p} = \left( \sum_{|\alpha| \le m} \|\partial^\alpha f\|_{L^p(\Omega)}^p \right)^{\frac{1}{p}} \qquad \text{if} \qquad 1 \le p < \infty \qquad (1.12.1)$$

or

$$\|f\|_{m,\infty} = \max_{|\alpha| \le m} \|\partial^\alpha f\|_{L^\infty(\Omega)} \qquad \text{if} \qquad p = \infty \qquad (1.12.2)$$

are called Sobolev spaces over $\Omega$.

Clearly, $W^{0,p}(\Omega) = L^p(\Omega)$. We first prove the following

THEOREM 1.12.1. $W^{m,p}(\Omega)$ is a Banach space.

PROOF. Let $(f_\nu)$ be a Cauchy sequence in $W^{m,p}(\Omega)$. Then for each $|\alpha| \le m$ there exists a function $f_\alpha$ in $L^p(\Omega)$ such that $\|\partial^\alpha f_\nu - f_\alpha\|_{L^p(\Omega)}$ tends to zero as $\nu \to \infty$. Since the sequence $(f_\nu)$ converges to $f$ in $L^p(\Omega)$, therefore the sequence

$(\partial^\alpha f_\nu)$ converges to $D^\alpha f$ in the distributional sense for each $\alpha \in \mathbb{N}^n$. Therefore $D^\alpha f = f_\alpha$ in $\mathcal{D}'(\Omega)$ if $|\alpha| \le m$. This means that

$$(-1)^{|\alpha|} \int_\Omega f \partial^\alpha \varphi = \int_\Omega f_\alpha \varphi$$

for $\varphi \in \mathcal{D}(\Omega)$. Therefore $f_\alpha$ is the weak derivative $\partial^\alpha f$ of $f$ if $|\alpha| \le m$. Thus, the proof is finished.                                                            $\square$

In the next chapter we shall show that $C^\infty(\Omega)$ is dense in $W^{m,p}(\Omega)$ if $\Omega$ is an arbitrary open set and $1 \le p < \infty$. In this section we restrict ourselves to presenting the fact that $\mathcal{D}(\Omega)$ is dense in $W^{m,p}(\Omega)$, when $\Omega = \mathbb{R}^n$ and $1 \le p < \infty$. Moreover, we show that $\mathcal{D}(\Omega)$ is not dense in $W^{m,p}(\Omega)$ if $\Omega \ne \mathbb{R}^n$ in general.

THEOREM 1.12.2. *If $1 \le p < \infty$, then $\mathcal{D}$ is dense in $W^{m,p}$.*

PROOF. Let $\psi$ be as in the previous section and $f \in W^{m,p}$. Then the regularization $f_\epsilon = \psi_\epsilon * f$ is a smooth function and $\partial^\alpha f_\epsilon = \partial^\alpha \psi_\epsilon * f$. By virtue of Young's inequality, it follows that $\partial^\alpha f_\epsilon$ is in $L^p$ for each $\alpha \in \mathbb{N}^n$.

Since $f$ is in $W^{m,p}$, therefore the weak derivatives $\partial^\alpha f$ of $f$ belong to $L^p$ for $|\alpha| \le m$. Taking into account the definition of weak derivative, by Theorem 1.3.1 we get

$$\frac{\partial^{|\alpha|}}{\partial x^\alpha} f_{\frac{1}{\nu}}(x) = \int_{\mathbb{R}^n} \psi_{\frac{1}{\nu}}(x - y) \partial^\alpha f(y) dy \qquad \text{for} \qquad \nu \in \mathbb{N}.$$

Moreover, from this theorem it also follows that $\partial^\alpha f_{\frac{1}{\nu}}$ tends to $\partial^\alpha f$ as $\nu \to \infty$ in $L^p$. This means that $C^\infty$ is dense in $W^{m,p}$.

What is left to show is that the following lemma holds.                              $\square$

LEMMA 1.12.1. *If $f$ is smooth and $\partial^\alpha f \in L^p$, $|\alpha| \le m$, then there exists a sequence $(f_\nu)$, $f_\nu \in \mathcal{D}$ such that $\partial^\alpha f_\nu$ tends to $\partial^\alpha f$ in $L^p$ for $|\alpha| \le m$.*

PROOF. Let $\chi_{B_\nu}$ be the characteristic function of the ball $B_\nu(0)$, $\nu \in \mathbb{N}$. Put

$$g_\nu(x) = (\chi_{B_\nu} * \psi)(x) = \int_{\mathbb{R}^n} \chi_{B_\nu}(x - y) \psi(y) \, dy.$$

Note that $g_\nu(x) = 1$ for $x \in B_{\nu-1}(0)$. Assume now that $f$ is a smooth function and $\partial^\alpha f$ belongs to $L^p$ for $|\alpha| \le m$. We shall now show that $\partial^\alpha(g_\nu f)$ tends to $\partial^\alpha f$ as $\nu \to \infty$ in $L^p$. Of course,

$$\partial^\alpha(g_\nu f) = g_\nu \partial^\alpha f + \sum_{0 < \beta \le \alpha} \binom{\alpha}{\beta} \partial^\beta g_\nu \partial^{\alpha-\beta} f.$$

It is easy to check that

$$\frac{\partial^{|\beta|}}{\partial x^\beta} g_\nu(x) = 0 \qquad \text{for} \qquad x \in B_{\nu-1}(0) \text{ and } \beta > 0.$$

Therefore

$$\frac{\partial^{|\alpha|}}{\partial x^\alpha}(g_\nu f)(x) = \frac{\partial^{|\alpha|}}{\partial x^\alpha} f(x) \qquad \text{for} \qquad x \in B_{\nu-1}(0).$$

It is easily seen that

$$\|\partial^\beta g_\nu\|_{L^\infty} \le \|\partial^\beta \psi\|_{L^1} =: M_\beta.$$

Hence we have

$$\left(\int_{\mathbb{R}^n} |\partial^\alpha(g_\nu f) - \partial^\alpha f|^p\right)^{\frac{1}{p}} \le \left(\int_{\mathbb{R}^n - B_{\nu-1}} \left|\sum_{0<\beta\le\alpha} \binom{\alpha}{\beta}\partial^\beta g_\nu \partial^{\alpha-\beta} f\right|^p\right)^{\frac{1}{p}}$$

$$\le \sum_{0<\beta\le\alpha} \binom{\alpha}{\beta} M_\beta \|\partial^{\alpha-\beta} f\|_{L^p(\mathbb{R}^n - B_{\nu-1})}.$$

Since $\partial^\alpha f \in L^p$ for $|\alpha| \le m$, therefore the right side of the above inequality is small for large $\nu$. This statement finishes the proof. $\square$

EXAMPLE 1.12.1. We shall now show that $\mathcal{D}(0,1)$ is not dense in $W^{1,2}(0,1)$. Indeed, let $\varphi$ be in $\mathcal{D}(0,1)$, then

$$\|\varphi\|_{L^2(0,1)} = \left(\int_0^1 \left|\int_0^x \varphi'(y)dy\right|^2 dx\right)^{\frac{1}{2}} \le \left(\int_0^1 \left(\int_0^1 |\varphi'(y)| dy\right)^2 dx\right)^{\frac{1}{2}}$$

$$\le \left(\int_0^1 \|\varphi'\|_{L^2(0,1)}^2 dx\right)^{\frac{1}{2}} \le \|\varphi'\|_{L^2(0,1)}.$$

Therefore

$$\|\varphi\|_{L^2(0,1)} \le \|\varphi'\|_{L^2(0,1)} \tag{1.12.3}$$

for $\varphi \in \mathcal{D}(0,1)$. Suppose that for some $\varphi \in \mathcal{D}(0,1)$, $\|\varphi - \chi_{(0,1)}\|_{1,2} < \frac{1}{2}$, then

$$\int_0^1 |\varphi - \chi_{(0,1)}|^2 + \int_0^1 |\varphi'|^2 < \frac{1}{4}.$$

Hence $\|\varphi'\|_{L^2(0,1)} < \frac{1}{2}$. Consequently

$$\|\varphi'\|_{L^2(0,1)} < 1 - \|\varphi - \chi_{(0,1)}\|_{1,2} \le \|\varphi\|_{L^2(0,1)}.$$

This statement gives us a contradiction with (1.12.3).

Let $W_0^{m,p}(\Omega)$ denote the closure of $\mathcal{D}(\Omega)$ in the space $W^{m,p}(\Omega)$. The above example shows that $W_0^{m,p}\mathcal{D}(\Omega)$ is different from $W^{m,p}(\Omega)$ in general.

## 1.13. Differential equations of the second order with measures as coefficients

We shall consider the boundary value problem

$$-u'' + \mu_1 u = \mu_2$$
$$u(a) = u(b) = 0, \tag{1.13.1}$$

where $\mu_1$ and $\mu_2$ are real Borel measures, $\mu_1 \ge 0$. If $\mu_1$ and $\mu_2$ are integrable functions with respect to the Lebesgue measure, then the Ritz–Galerkin method is often used to investigate Problem (1.13.1). Here we shall show that this method may also be applied to solving Problem (1.13.1) under the above assumptions. We are looking for a continuous function $u$ vanishing at the end points $a, b$ and fulfilling equation (1.13.1) in the distributional sense. This means that

$$-\int_a^b u\varphi'' \, dx + \int_a^b u\varphi \, d\mu_1(x) = \int_a^b \varphi \, d\mu_2(x) \quad \text{for} \quad \varphi \in \mathcal{D}(a, b), \qquad (1.13.2)$$

where $\int_a^b \varphi \, d\mu_2(x) := \int_a^b \varphi q \, d|\mu_2|(x)$, $|\mu_2|$ is the variation of $\mu_2$ and $|q(x)| = 1$ a.e. (see Section 2.7). One can show that

$$W_0^{1,2}(a, b) = \left\{ u : u \text{ is absolutely continuous, } u(a) = u(b) = 0 \text{ and } u' \in L^2(a, b) \right\}.$$

For simplicity of notation we put

$$\alpha(\varphi, \psi) := \int_a^b \varphi' \psi' \, dx + \int_a^b \varphi \psi \, d\mu_1(x),$$

$$\beta(\varphi) := \int_a^b \varphi \, d\mu_2(x)$$

for $\varphi, \psi \in W_0^{1,2}(a, b)$. It is easy to check that $\alpha$ is a bilinear symmetric positive definite form on $W_0^{1,2}(a, b)$ and $\beta$ is a linear form on $W_0^{1,2}(a, b)$. If we are looking for a solution $u$ in $W_0^{1,2}(a, b)$, then equation (1.13.2) is equivalent to the equation

$$\int_a^b u' \varphi' \, dx + \int_a^b u\varphi \, d\mu_1(x) = \int_a^b \varphi \, d\mu_2(x), \qquad \varphi \in \mathcal{D}(a, b).$$

or, using the above notation, also

$$\alpha(u, \varphi) = \beta(\varphi), \qquad \varphi \in \mathcal{D}(a, b). \qquad (1.13.3)$$

In the sequel we shall need the following two norms

$$\|u\|_\alpha := [\alpha(u, u)]^{\frac{1}{2}} \quad \text{and} \quad \|u\|_D := \|u'\|_{L^2}$$

for $u$ in $W_0^{1,2}(a, b)$. Now we are in a position to state

THEOREM 1.13.1. *The following norms* $\| \cdot \|_{1,2}$, $\| \cdot \|_D$ *and* $\| \cdot \|_\alpha$ *are equivalent on* $W_0^{1,2}(a, b)$.

PROOF. Similarly to Example 1.12.1 one can show that

$$\|\varphi\|_{L^2} \leq (b - a)\|\varphi\|_D. \qquad (1.13.4)$$

It implies that the norms $\| \cdot \|_{1,2}$ and $\| \cdot \|_D$ are equivalent on the space $W_0^{1,2}(a, b)$. It is easy to see that

$$|\varphi(x)| \leq (b - a)^{\frac{1}{2}}\|\varphi\|_D$$

for $x \in [a, b]$ and $\varphi \in W_0^{1,2}(a, b)$. Therefore

$$\|\varphi\|_{L^\infty} \leq (b - a)^{\frac{1}{2}}\|\varphi\|_D. \qquad (1.13.5)$$

Note that
$$\|\varphi\|_\alpha \leq \|\varphi\|_D + \|\varphi\|_{L^2\mu_1}$$
for $\varphi \in W_0^{1,2}(a,b)$. Therefore we have
$$\|\varphi\|_D \leq \|\varphi\|_\alpha \leq \|\varphi\|_D + \|\varphi\|_{L^2\mu_1}$$
for $\varphi \in W_0^{1,2}(a,b)$. By (1.13.5) we obtain
$$\|\varphi\|_{L^2_{\mu_1}}^2 \leq \|\varphi\|_{L^\infty}^2 \mu_1([a,b]) \leq (b-a)\|\varphi\|_D^2 \mu_1([a,b]).$$
Finally we get
$$\|\varphi\|_D \leq \|\varphi\|_\alpha \leq \left[1 + ((b-a)\mu_1([a,b]))^{\frac{1}{2}}\right]\|\varphi\|_D. \tag{1.13.6}$$
Thus, the proof of our theorem is finished. $\qquad\square$

Let $\langle\cdot,\cdot\rangle_{L^2}$ and $\langle\cdot,\cdot\rangle_{L^2_{\mu_1}}$ denote the ordinary inner product on the space $L^2(a,b)$ and $L^2_{\mu_1}(a,b)$. We set
$$\langle\varphi,\psi\rangle_D := \langle\varphi',\psi'\rangle_{L^2}, \qquad\qquad \varphi,\psi \in W_0^{1,2}(a,b),$$
$$\langle\varphi,\psi\rangle := \langle\varphi,\psi\rangle_D + \langle\varphi,\psi\rangle_{L^2}, \qquad \varphi,\psi \in W_0^{1,2}(a,b),$$
and thus we have
$$\alpha(\varphi,\psi) = \langle\varphi,\psi\rangle_D + \langle\varphi,\psi\rangle_{L^2_{\mu_1}}$$
for $\varphi,\psi \in W_0^{1,2}(a,b)$. We know that $(W_0^{1,2}(a,b),\langle\cdot,\cdot\rangle)$ is a Hilbert space.

COROLLARY 1.13.1. *The spaces $(W_0^{1,2}(a,b),\langle\cdot,\cdot\rangle_D)$ and $(W_0^{1,2}(a,b),\alpha(\cdot,\cdot))$ are Hilbert spaces, too.*

Now, we are in a position to prove the main

THEOREM 1.13.2. *Problem (1.13.1) has exactly one solution in $W_0^{1,2}(a,b)$.*

PROOF. By the definition of $W_0^{1,2}(a,b)$, the set $\mathcal{D}(a,b)$ is dense in $W_0^{1,2}(a,b)$ so there exists at most one solution of Problem (1.13.1) in $W_0^{1,2}(a,b)$. Since the space $(W_0^{1,2}(a,b),\langle\cdot,\cdot\rangle)$ is a Hilbert space and $\beta$ is a continuous linear form on $W_0^{1,2}(a,b)$, there exists a function $u$ in $W_0^{1,2}(a,b)$ such that (1.13.3) holds. This finishes the proof. $\qquad\square$

In general, there exist no more regular solutions of Problem (1.13.1), apart from those belonging to $W_0^{1,2}(a,b)$.

EXAMPLE 1.13.1. Let us consider the differential equation
$$-x'' + \delta_{\frac{1}{2}}x = f$$
with the boundary condition
$$x(0) = x(1) = 0,$$
where $\delta_{\frac{1}{2}}$ is the Dirac measure concentrated at the point $t = \frac{1}{2}$ and $f \in L^1(0,1)$. It is easy to see that this problem has no classical solutions (belonging to $W_0^{2,2}(0,1)$).

Now, we use the Ritz–Galerkin method to determine approximate solutions of Problem (1.13.1). We begin with a formulation of the Ritz theorem.

Let $E$ be a real vector space and $\alpha : E \times E \to \mathbb{R}$ be a bilinear symmetric positive definite form. Moreover, let $\beta : E \to \mathbb{R}$ be a linear form. Let us consider the quadratic form

$$F(x) := \frac{1}{2}\alpha(x, x) - \beta(x).$$

THEOREM 1.13.3 (Ritz). *The following conditions are equivalent:*

(i) $\alpha(x, y) = \beta(y)$ *for* $y \in E$,

(ii) $F(x) = \inf_{y \in E} F(y)$.

We introduce the norm $\|x\|_\alpha = [\alpha(x, x)]^{\frac{1}{2}}$ in the space $E$. If we assume that $(E, \|\cdot\|_\alpha)$ is complete, then $(E, \alpha(\cdot, \cdot))$ is a Hilbert space. Let $(x_n)$, $n = 1, 2, \ldots$ be a sequence of elements $x_n$ belonging to $E$ such that

$$\overline{\operatorname{lin}\{x_n : \ n = 1, 2, \ldots\}} = E, \tag{1.13.7}$$

where $\operatorname{lin}\{x_n : \ n = 1, 2, \ldots\}$ denotes the vector space spanned by the elements $x_n$. Let $E_n$ be the space spanned by elements $x_1 \ldots x_n$. Let $x_n^*$ be an element in $E_n$ such that $F(x_n^*) = \inf_{y \in E_n} F(y)$. It is known that $\|x_n^* - x\|_\alpha$ tends to zero, when $F(x) = \inf_{y \in E} F(y)$.

The above information will be used in determining approximate solutions of Problem (1.13.1) in the space $W_0^{1,2}(a, b)$. Let $E_n$ be the vector space spanned by the functions $f_2, \ldots, f_n$. The quadratic form $F$ takes the following form

$$F(y) = G(\lambda_2, \ldots, \lambda_n)$$

$$= \frac{1}{2}\left(\sum_{i=2}^{n}\sum_{j=2}^{n}\lambda_i\lambda_j\int_a^b f_i' f_j' \, dx + \sum_{i=2}^{n}\sum_{j=2}^{n}\lambda_i\lambda_j\int_a^b f_i f_j \, d\mu_1(x)\right) - \sum_{i=2}^{n}\lambda_i\int_a^b f_i \, d\mu_2(x), \tag{1.13.8}$$

where $y = \lambda_2 f_2 + \ldots + \lambda_n f_n$. Formula (1.13.8) may be rewritten in matrix form

$$G(\Lambda) = \frac{1}{2}\Lambda^T(\Gamma + \Delta)\Lambda - \Lambda^T P,$$

where

$$\Lambda = \begin{bmatrix} \lambda_2 \\ \vdots \\ \lambda_n \end{bmatrix}, \qquad \Gamma = \begin{bmatrix} \langle f_2, f_2 \rangle_D & \cdots & \langle f_2, f_n \rangle_D \\ \vdots & & \vdots \\ \langle f_n, f_2 \rangle_D & \cdots & \langle f_n, f_n \rangle_D \end{bmatrix},$$

$$\Delta = \begin{bmatrix} \int_a^b f_2 f_2 \, d\mu_1(x) & \cdots & \int_a^b f_2 f_n \, d\mu_1(x) \\ \vdots & & \vdots \\ \int_a^b f_n f_2 \, d\mu_1(x) & \cdots & \int_a^b f_n f_n \, d\mu_1(x) \end{bmatrix}, \qquad P = \begin{bmatrix} \int_a^b f_2 \, d\mu_2(x) \\ \vdots \\ \int_a^b f_n \, d\mu_2(x) \end{bmatrix}.$$

It is easy to check that

$$G(\lambda^*) = \inf_{\lambda \in \mathbb{R}^{n-1}} G(\lambda),$$

when

$$(\Gamma + \Delta)\lambda^* = P. \tag{1.13.9}$$

Now, we construct a sequence $(f_n)$ in the space $W_0^{1,2}(0,1)$ which has property (1.13.7). For $m = 2$ we put $f_2(t) := t$ for $0 \leq t \leq \frac{1}{2}$, $f_2(t) = 1 - t$ for $\frac{1}{2} < t \leq 1$. In the general case we take

$$f_m(t) := \begin{cases} 2^{\frac{n}{2}}t - \frac{2k-2}{2^{\frac{n}{2}+1}} & \text{for } \frac{2k-2}{2^{n+1}} \leq t < \frac{2k-1}{2^{n+1}} \\ -2^{\frac{n}{2}}t + \frac{2k}{2^{\frac{n}{2}+1}} & \text{for } \frac{2k-1}{2^{n+1}} \leq t \leq \frac{2k}{2^{n+1}} \\ 0 & \text{for other } t \text{ in } [0,1], \end{cases}$$

where $m = 2^n + k$, $1 \leq k \leq 2^n$, $n = 0, 1, 2, \ldots$

THEOREM 1.13.4. *The functions $f_n$, $n = 2, 3, \ldots$ constitute a complete orthonormal system in $(W_0^{1,2}(0,1), \langle \cdot, \cdot \rangle_D)$.*

PROOF. Note that the weak derivatives of the functions $f_n$ are known Haar functions $\chi_n$. For $f \in W_0^{1,2}(0,1)$ we have

$$f(x) = \int_0^x g(t)\, dt$$

for some $g \in L^2(0,1)$, $x \in [0,1]$. The function $g$ has the Fourier representation

$$g = \int_0^1 g(x)\, dx + \sum_{n=2}^{\infty} c_n \chi_n \tag{1.13.10}$$

with respect to the Haar functions. It is clear that $\int_0^1 g(x)\, dx = 0$. Since the space $(W_0^{1,2}(0,1), \langle \cdot, \cdot \rangle_D)$ is complete, therefore there exists a function $\tilde{f}$ in $W_0^{1,2}(0,1)$ such that

$$\sum_{n=2}^{\infty} c_n f_n = \tilde{f}. \tag{1.13.11}$$

Series (1.13.11) converges in the distributional sense. Hence we have

$$\sum_{n=2}^{\infty} c_n \chi_n = \tilde{f}'. \tag{1.13.12}$$

It is known that series (1.13.12) converges to $\tilde{f}'$ in the space $L^2(0,1)$ (also a.e.). This implies that $\tilde{f}' = g$ a.e. on $[0,1]$. From this we obtain that $f(x) = \tilde{f}(x)$ for each $x \in [0,1]$. Finally we have

$$f(x) = \sum_{n=2}^{\infty} \langle f, f_n \rangle_D f_n.$$

This completes the proof of the theorem. □

COROLLARY 1.13.2. $\overline{\text{lin}\{f_n : n = 2, 3, \ldots\}} = W_0^{1,2}(0,1)$ *with respect to the norm $\|\cdot\|_\alpha$.*

For the differential equation

$$-x'' + \delta_{\frac{1}{2}} x = 1$$
$$x(0) = x(1) = 0 \tag{1.13.13}$$

we obtain the following matrix equation

$$\begin{bmatrix} \frac{5}{4} & 0 & \cdots & \cdots & 0 \\ 0 & 1 & & & \vdots \\ \vdots & 0 & \ddots & & \vdots \\ \vdots & & \ddots & \ddots & 0 \\ 0 & \cdots & \cdots & 0 & 1 \end{bmatrix} \begin{bmatrix} \lambda_2 \\ \vdots \\ \vdots \\ \vdots \\ \lambda_{2m+1} \end{bmatrix} = \begin{bmatrix} a_2 \\ \vdots \\ \vdots \\ \vdots \\ a_{2m+1} \end{bmatrix},$$

where $a_2 = \frac{1}{4}$, $a_{2^k+l} = 2^{-\frac{3k}{2}-2}$ for $l = 1, 2, \ldots, 2^k$, $k = 1, \ldots, m$. The exact solution of (1.13.13) is

$$x(t) = \begin{cases} -\frac{t^2}{2} + \frac{9}{20}t & \text{for } 0 \le t \le \frac{1}{2} \\ -\frac{t^2}{2} + \frac{11}{20}t - \frac{1}{20} & \text{for } \frac{1}{2} < t \le 1. \end{cases}$$

The following graphs compare $x$ and $x_n^*$ for $n = 2, 4, 8$.

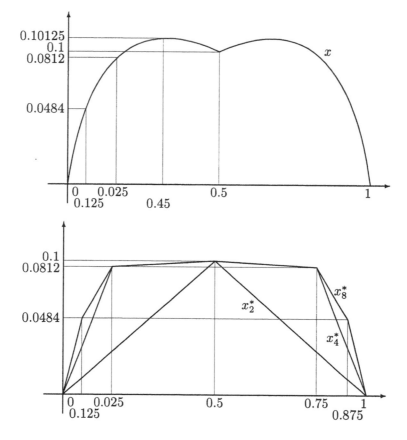

## CHAPTER 2

# LOCAL PROPERTIES OF DISTRIBUTIONS

## 2.1. Smooth partitions of unity

We have already seen that distributions cannot be assigned values at isolated points. One cannot, for instance, say that "a distribution $\Lambda$ is equal to zero at $x_0$". However, the statement that "$\Lambda$ is equal to zero in neighbourhood $U$ of $x_0$" can be meaningful. This will mean, namely, that for every $\varphi \in \mathcal{D}(\Omega)$ with support in $U$, we have $\Lambda(\varphi) = 0$. For example, the Dirac measure is equal to zero in the neighbourhood of every point $x \neq 0$. We say that a distribution $\Lambda \in \mathcal{D}'(\Omega)$ vanishes in an open set $\Omega' \subset \Omega$ if for every $\varphi \in \mathcal{D}(\Omega)$ with support in $\Omega'$ we have $\Lambda(\varphi) = 0$. A typically local property is the following one. A distribution vanishes in an open set $\Omega' \subset \Omega$ iff it equals zero in the neighbourhood of every point in this open set. This fact can be used to construct a distribution in its entirely when it is defined everywhere only locally. We shall prove this property in Section 2.3.

Now, we start with presenting the following

LEMMA 2.1.1. *Let $\Omega$ be an open set in $\mathbb{R}^n$ and $K \subset \Omega$ be a compact set. Then there exists a function $f$ in $\mathcal{D}(\Omega)$ such that $0 \leq f(x) \leq 1$ for $x \in \Omega$ and $f(x) = 1$ for $x \in K$.*

PROOF. Take a function $\psi$ in $\mathcal{D}$ such that $\psi \geq 0$, $\operatorname{supp} \psi \subset B_1(0)$ and $\int \psi = 1$ and put

$$\psi_\epsilon(x) := \epsilon^{-n} \psi \left( \frac{x}{\epsilon} \right), \quad \epsilon > 0.$$

Further, assume $2\epsilon < \operatorname{dist}(K, \partial\Omega)$. The characteristic function $\chi_{\overline{B_\epsilon(K)}}$ of $\overline{B_\epsilon(K)}$ is integrable. We shall now show that the regularization $\chi_\epsilon$,

$$\chi_\epsilon(x) - \int_\Omega \psi_\epsilon(x - y) \chi_{\overline{B_\epsilon(K)}}(y) \, dy$$

may be taken as a required function $f$. Indeed, the function $\chi_\epsilon$ is smooth (see Section 1.11). An easy computation shows that $\operatorname{supp} \chi_\epsilon \subset \Omega$, $\chi_\epsilon(x) = 1$ for $x \in K$ and $0 \leq \chi_\epsilon(x) \leq 1$ for $x \in \Omega$. Thus, the proof is finished. $\square$

The following theorem is important for applications.

THEOREM 2.1.1. *Let $\vartheta = \{\Omega_\alpha : \alpha \in A\}$ be a collection of open sets. Assume that $\varphi \in \mathcal{D}(\Omega)$ and $\Omega \subset \bigcup_{\alpha \in A} \Omega_\alpha$, then there exists a finite subcollection $\Omega_1, \ldots, \Omega_k$ of $\vartheta$ and a finite collection of functions $\varphi_\nu$, $\varphi_\nu \in \mathcal{D}(\Omega_\nu)$, $\nu = 1, \ldots, k$ such that*

$$\varphi = \sum_{\nu=1}^{k} \varphi_\nu. \tag{2.1.1}$$

PROOF. Since supp $\varphi$ is compact, therefore there exists a finite subcollection $\Omega_1, \ldots, \Omega_m \subset \vartheta$ such that supp $\varphi \subset \bigcup_{j=1}^m \Omega_j$. Let $x \in \text{supp}\,\varphi$, then there exists an open neighbourhood $U_x$ of $x$ contained in $\Omega_j$ for some $j$. Obviously, we choose $U_x$ so that $\overline{U}_x \subset \Omega_j$. Because of the compactness of supp $\varphi$ one can choose a finite set of the neighbourhoods $U_{x_1}, \ldots, U_{x_n}$, $x_i \in \text{supp}\,\varphi$ so that supp $\varphi \subset \bigcup_{i=1}^n U_{x_i}$ and $U_{x_i} \subset \Omega_j$ for some $j$. Put $K_j := \bigcup_{i=1}^{m_j} \overline{U}_{x_i}$. In view of Lemma 2.1.1 there exists a function $\psi_j \in \mathcal{D}(\Omega_j)$ such that $0 \leq \psi_j(x) \leq 1$, $x \in \Omega_j$ and $\psi_j(x) = 1$ for $x \in K_j$. We shall now define new functions as follows:

$$\varphi_1 = \varphi\psi_1,$$
$$\varphi_2 = \varphi\psi_2(1 - \psi_1),$$
$$\cdots$$
$$\varphi_k = \varphi\psi_k(1 - \psi_1) \cdots (1 - \psi_{k-1}).$$

We are able to show that

$$\sum_{j=1}^k \varphi_j - \varphi = -\varphi \prod_{j=1}^k (1 - \psi_j). \tag{2.1.2}$$

Of course, (2.1.2) is true for $k = 1$. Suppose

$$\sum_{j=1}^i \varphi_j - \varphi = -\varphi \prod_{j=1}^i (1 - \psi_j).$$

Therefore

$$\sum_{j=1}^i \varphi_j - \varphi + \varphi_{i+1} = -\varphi \prod_{j=1}^i (1 - \psi_j) + \varphi_{i+1}.$$

The expression on the right hand of this equality can be written as follows

$$-\varphi \prod_{j=1}^i (1 - \psi_j) + \varphi\psi_{i+1} \prod_{j=1}^i (1 - \psi_j) = -\varphi \prod_{j=1}^{i+1} (1 - \psi_j).$$

Thus, (2.1.2) is shown. Note that $\varphi(x) \prod_{j=1}^k (1 - \psi_j(x)) = 0$ for $x \in \Omega$. Indeed, if $x \in \text{supp}\,\varphi$, then $1 - \psi_j(x) = 0$ for some $j$ but if $x \notin \text{supp}\,\varphi$ then $\varphi(x) = 0$. Thus, we have proved (2.1.1). $\qquad\square$

DEFINITION 2.1.1. A collection $\vartheta = \{\Omega_\alpha : \alpha \in A\}$ of open sets $\Omega_\alpha \subset \mathbb{R}^n$ such that $A \subset \bigcup_{\alpha \in A} \Omega_\alpha$ is called an open covering of $A$.

THEOREM 2.1.2 (Smooth partition of unity). *Let $A$ be an arbitrary set in $\mathbb{R}^n$ and $\vartheta$ be an open covering of $A$. Then there exists a family $\Psi$ of functions $\psi$ in $\mathcal{D}$ which have the following properties:*

   (i) *$0 \leq \psi(x) \leq 1$ for $x \in \mathbb{R}^n$,*
   (ii) *if $K \Subset A$, then all except for possibly a large finite number of $\psi \in \Psi$ vanish identically on $K$,*
   (iii) *for each function $\psi$ in $\Psi$ there exists $\Omega \in \vartheta$ such that supp $\psi \subset \Omega$,*
   (iv) *for every $x \in A$, $\sum_{\psi \in \Psi} \psi(x) = 1$.*

DEFINITION 2.1.2. A collection $\Psi$ of functions $\psi$ from $\mathcal{D}$ having the properties listed in the above theorem is called a smooth partition of unity for $A$ subordinate to $\vartheta$.

PROOF OF THEOREM 2.1.2. The proof falls naturally into three parts. At the beginning, suppose that $A$ is a compact set. Let $\vartheta$ be an open covering of $A$. Then there exist sets $\Omega_i \in \vartheta$, $i = 1, \ldots, k$ such that $A \subset \bigcup_{i=1}^{k} \Omega_i := \Omega$. By virtue of Lemma 2.1.1 there exists a function $\psi$ in $D(\Omega)$ such that $\psi(x) = 1$ for $x \in A$. Taking into account Theorem 2.1.1 we infer that there exist functions $\psi_i$ in $\mathcal{D}(\Omega_i)$ such that $\psi = \sum_{i=1}^{k} \psi_i$. Thus, the proof is finished if $A$ is a compact set.

We shall now consider the case, when $A$ is an open set. Put

$$A_j = \left\{ x \in A : j - 1 \leq |x| \leq j, \ \mathrm{dist}(x, \partial A) \geq \frac{1}{j} \right\}$$

for $j = 1, 2, \ldots$ and $A_{-1} = A_0 = \emptyset$, the empty set. Then $A_j$ is compact and $A = \bigcup_{j=1}^{\infty} A_j$. For each $j$, the collection

$$\vartheta_j := \{ U \cap \mathrm{interior}\ A_{j+1} \cap (A_{j-2})^c : U \in \vartheta \}$$

covers $A_j$. In accordance with the above statement, there exist functions $\Theta_{j,1} \ldots \Theta_{j,k_j}$ which are subordinated to the covering $\vartheta_j$ and $\sum_{\nu=1}^{k_j} \Theta_{j\nu}(x) = 1$ for $x \in A_j$. Put

$$\sigma(x) := \sum_{j=1}^{\infty} \sum_{\nu=1}^{k_j} \Theta_{j\nu}(x).$$

Of course, $\sigma(x) > 0$ for $x \in A$. For each $x \in A$ the above sum consists of only a finite number of summands. Let

$$\Psi = \left\{ \psi_{j\nu} : \psi_{j\nu} = \frac{\Theta_{j\nu}}{\sigma}, \ j = 1, 2, \ldots, \ \nu = 1 \ldots k_j \right\}.$$

Note that the collection $\Psi$ has the properties (i), (ii), (iii) and (iv).

Finally, if $A$ is an arbitrary set, then $A \subset B$, where $B$ is open. Any partition of unity for $B$ will do for $A$, as well. $\qquad\square$

## 2.2. Approximation of functions belonging to $W^{m,p}(\Omega)$ by smooth functions

In Section 1.12 it was proved that $\mathcal{D}(\Omega)$ is dense in $W^{m,p}(\Omega)$ if $\Omega = \mathbb{R}^n$ and $1 \leq p < \infty$. Example 1.12.1 indicates that $D(\Omega)$ is not dense in $W^{m,p}(\Omega)$ in general. We shall show that $C^{\infty}(\Omega)$ is dense in $W^{m,p}(\Omega)$ for any open set $\Omega$ and $1 \leq p < \infty$.

THEOREM 2.2.1 (Meyers–Seerin). $C^{\infty}(\Omega)$ is dense in $W^{m,p}(\Omega)$ for any open set $\Omega$ and $1 \leq p < \infty$.

PROOF. Let $f$ be in $W^{m,p}(\Omega)$ and $\epsilon > 0$, we have to show that there exists $\Phi \in C^{\infty}(\Omega)$ such that $\|f - \Phi\|_{m,p,\Omega} < \epsilon$. Consider the collection of the open sets

$$\Omega_k = \left\{ x \in \Omega : \ |x| < k \ \text{and} \ \mathrm{dist}(x, \partial\Omega) > \frac{1}{k} \right\}$$

for $k = 1, 2, \ldots$. For $k = -1, 0$ we take $\Omega_{-1} = \Omega_0 = \emptyset$. The collection

$$\vartheta = \{ V_k : \ V_k = \Omega_{k+1} \cap (\overline{\Omega}_{k-1})^c, \ k = 1, 2, \ldots \}$$

of open sets $V_k$ covers $\Omega$. Let $\Psi$ be a smooth partition of unity for $\Omega$ subordinate to $\vartheta$. By $\psi_k$, we denote the sum of a large finite number of functions $\psi$, $\psi \in \Psi$ whose supports are contained in $V_k$. Then $\psi_k \in \mathcal{D}(V_k)$ and $\sum_{k=1}^{\infty} \psi_k(x) = 1$ for $x \in \Omega$. Since $\psi_k f \in W^{m,p}(\Omega)$, by Lemma 1.11.1 we have

$$\frac{\partial^{|\alpha|}}{\partial x^\alpha}(\psi_k f)_\epsilon(x) = (\partial^\alpha(\psi_k f))_\epsilon(x)$$

if $x \in \Omega$ and $\operatorname{dist}(x, \partial\Omega) > \epsilon$. Note that

$$\operatorname{supp}(\psi_k f)_\epsilon \subset \Omega_{k+2} \cap (\Omega_{k-2})^c = V_k \Subset \Omega$$

if $0 < \epsilon < \frac{1}{k+1} \cdot \frac{1}{k+2} = \frac{1}{k+1} - \frac{1}{k+2}$. For $\epsilon > 0$ in view of Lemma 1.11.2 we can choose $\epsilon_k$, $0 < \epsilon_k < \frac{1}{k+1} \cdot \frac{1}{k+2}$ so that

$$\|(\partial^\alpha(\psi_k f))_{\epsilon_k} - \partial^\alpha(\psi_k f)\|_{L^2(V_k)} < \frac{\epsilon}{2^k}.$$

Since $\operatorname{dist}(V_k, \partial\Omega) < \epsilon$ and $\operatorname{supp}((\psi_k f)_\epsilon - \psi_k f) \subset V_k$, therefore

$$\|\partial^\alpha(\psi_k f)_{\epsilon_k} - \partial^\alpha(\psi_k f)\|_{L^p(\Omega)} = \|(\partial^\alpha(\psi_k f))_{\epsilon_k} - \partial^\alpha(\psi_k f)\|_{L^p(\Omega)}$$
$$= \|\partial^\alpha(\psi_k f)_{\epsilon_k} - \partial^\alpha(\psi_k f)\|_{L^p(V_k)} < \frac{\epsilon}{2^k}$$

if $|\alpha| \leq m$. We can choose $\epsilon_k$ such that

$$\|(\psi_k f)_{\epsilon_k} - \psi_k f\|_{m,p,\Omega} = \|(\psi_k f)_{\epsilon_k} - \psi_k f\|_{m,p,V_k} < \frac{\epsilon}{2^k} \qquad (2.2.1)$$

because

$$\|(\psi_k f)_{\epsilon_k} - \psi_k f\|_{m,p,V_k} = \max_{|\alpha| \leq m} \|\partial^\alpha(\psi_k f)_{\epsilon_k} - \partial^\alpha(\psi_k f)\|_{L^p(V_k)}.$$

Let $\Phi = \sum_{k=1}^{\infty}(\psi_k f)_{\epsilon_k}$. On any $\Omega' \Subset \Omega$ this sum involves only a large finite number of nonzero terms. Therefore $\Phi$ is in $C^\infty(\Omega)$. For $x \in \Omega_k$ we have

$$f(x) = \sum_{j=1}^{k+2} \psi_j(x) f(x) \quad \text{and} \quad \Phi(x) = \sum_{j=1}^{k+2}(\psi_j f)_{\epsilon_j}(x).$$

Thus

$$\|f - \Phi\|_{m,p,\Omega_k} \leq \sum_{j=1}^{k+2} \|\psi_j f - (\psi_j f)_{\epsilon_j}\|_{m,p,\Omega}.$$

Taking into account (2.2.1) we get

$$\|f - \Phi\|_{m,p,\Omega_k} < \epsilon.$$

Finally, by the Lebesgue monotone convergence theorem

$$\|f - \Phi\|_{m,p,\Omega} < \epsilon. \qquad \square$$

REMARK 2.2.1. Let $\Omega = (-1,1)$, $f(x) = |x|$. Of course, $f \in W^{1,\infty}(-1,1)$. Note that $f'(x) = -1$ for $x \in (-1,0)$ and $f'(x) = 1$ for $x \in (0,1)$. Obviously, there is no function $\Phi \in C^\infty(-1,1)$ such that $\|\Phi' - f'\|_{L^\infty(-1,1)} < \epsilon$ if $\epsilon < \frac{1}{2}$.

## 2.3. Restrictions of distributions

Let $\Omega'$ be an open subset of the open set $\Omega$. For $\Lambda \in \mathcal{D}'(\Omega)$ put

$$\Lambda|_{\Omega'}(\varphi) := \Lambda(\varphi) \text{ if } \varphi \in \mathcal{D}(\Omega').$$

Obviously $\Lambda|_{\Omega'}$ is a distribution on $\Omega'$. This distribution will be called the restriction of $\Lambda$ to the open set $\Omega'$.

THEOREM 2.3.1. *If $\Lambda \in \mathcal{D}'(\Omega)$ and for each $x \in \Omega$ there exists an open neighbourhood $U_x$ such that $\Lambda|_{U_x} = 0$, then $\Lambda = 0$ on $\Omega$.*

PROOF. Let $\varphi$ in $\mathcal{D}(\Omega)$ and $x \in \Omega$, then in accordance with the assumption there exists an open neighbourhood $U_x$ such that $\Lambda|_{U_x} = 0$. Since $\operatorname{supp} \varphi$ is a compact set, therefore one can choose a finite number of sets $U_{x_1}, \ldots, U_{x_k}$, $x_i \in \Omega$ so that $\operatorname{supp} \varphi \subset \bigcup_{i=1}^{k} U_{x_i}$. By Theorem 2.1.1 there exist functions $\varphi_i \in \mathcal{D}(U_{x_i})$ such that $\varphi = \sum_{i=1}^{k} \varphi_i$. Hence we obtain

$$\Lambda(\varphi) = \sum_{i=1}^{k} \Lambda|_{U_{x_i}}(\varphi_i) = 0. \qquad \square$$

We say that the distributions $\Lambda_1$ and $\Lambda_2$ coincide in a neighbourhood of $x_0$ if their difference $\Lambda_1 - \Lambda_2$ vanishes in this neighbourhood. We have proved that distributions which coincide in the neighbourhood of every point of $\Omega$ are equal to each other on $\Omega$. Thus, every distribution is uniquely determined by its local properties. This fact can be used to construct a distribution in its entirely on $\Omega$, when it is defined on $\Omega_\alpha$, $\alpha \in \mathcal{A}$ and $\Omega = \bigcup_{\alpha \in \mathcal{A}} \Omega_\alpha$.

THEOREM 2.3.2. *Let $\vartheta$ be a collection of open sets $\Omega_\alpha$, $\alpha \in \mathcal{A}$ and $\Omega = \bigcup_{\alpha \in \mathcal{A}} \Omega_\alpha$. Assume that for each $\alpha \in \mathcal{A}$ is given a distribution $\Lambda_\alpha \in \mathcal{D}'(\Omega_\alpha)$ such that for two indices $\alpha_1$ and $\alpha_2$, $\Lambda_{\alpha_1}(\varphi) = \Lambda_{\alpha_2}(\varphi)$ for $\varphi \in \mathcal{D}(\Omega_{\alpha_1} \cap \Omega_{\alpha_2})$, then there exists exactly one distribution $\Lambda \in \mathcal{D}'(\Omega)$ such that $\Lambda|_{\Omega_\alpha} = \Lambda_\alpha$ for every $\alpha \in \mathcal{A}$.*

PROOF OF UNIQUENESS. Suppose that there are two distributions $\Lambda_1$ and $\Lambda_2$ belonging to $D'(\Omega)$ satisfying our assumptions. Let $\operatorname{supp} \varphi \subset \bigcup_{i=1}^{k} \Omega_{\alpha_j}$. Because of Theorem 2.1.1 there exist functions $\varphi_j \in \mathcal{D}(\Omega_{\alpha_j})$ such that $\varphi = \sum_{j=1}^{k} \varphi_j$. Therefore

$$\Lambda_1(\varphi) = \sum_{j=1}^{k} \Lambda_{\alpha_j}(\varphi_j) = \Lambda_2(\varphi).$$

Thus, the proof of uniqueness is finished. $\qquad \square$

PROOF OF EXISTENCE. Put as in the above consideration

$$\Lambda(\varphi) = \sum_{j=1}^{k} \Lambda_{\alpha_j}(\varphi_j), \tag{2.3.1}$$

where $\varphi_j \in \mathcal{D}(\Omega_{\alpha_j})$. We have to show that $\Lambda(\varphi)$ does not depend on representations of $\varphi$. Take two representations $\varphi = \sum_{j=1}^{k_1} \varphi_j^1$ and $\varphi = \sum_{j=1}^{k_2} \varphi_j^2$. Without loss of generality we can assume that $k_1 = k_2 = k$. Note that

$$\Lambda(0) = \sum_{j=0}^{k} \Lambda_{\alpha_j}(\varphi_j^1 - \varphi_j^2) = \sum_{j=1}^{k} \Lambda_{\alpha_j}(\varphi_j),$$

where $\varphi_j = \varphi_j^1 - \varphi_j^2$. What is left to show is that the following condition

$$\sum_{j=1}^k \varphi_j = 0 \implies \sum_{j=1}^k \Lambda_{\alpha_j}(\varphi_j) = 0 \tag{2.3.2}$$

holds. Let $F = \bigcup_{j=1}^k \operatorname{supp} \varphi_j$, $\varphi_j \in \mathcal{D}(\Omega_{\alpha_j})$. By Lemma 2.1.1 there exists a function $\psi \in \mathcal{D}\left(\bigcup_{j=1}^k \Omega_{\alpha_j}\right)$ such that $\psi(x) = 1$ on $F$. The function $\psi$ can be written as follows $\psi = \sum_{j=1}^k \psi_j$, where $\psi_j \in \mathcal{D}(\Omega_{\alpha_j})$. We proceed to show that the implication (2.3.2) holds. Assume that $\sum_{j=1}^k \varphi_j = 0$. Since $F = \bigcup_{j=1}^k \operatorname{supp} \varphi_j$, therefore $\psi\varphi_j = \varphi_j$ for $j = 1, \ldots k$. Hence we have

$$\sum_{j=1}^k \Lambda_{\alpha_j}(\varphi_j) = \sum_{j=1}^k \Lambda_{\alpha_j}(\psi\varphi_j) = \sum_{j=1}^k \Lambda_{\alpha_j}\left(\sum_{i=1}^k \psi_i\varphi_j\right)$$
$$= \sum_{j=1}^k \sum_{i=1}^k \Lambda_{\alpha_j}(\psi_i\varphi_j).$$

Note that $\operatorname{supp} \psi_i\varphi_j \subset \Omega_i \cap \Omega_j$. In accordance with the assumption we have $\Lambda_{\alpha_i}(\psi_i\varphi_j) = \Lambda_{\alpha_j}(\psi_i\varphi_j)$, therefore

$$\sum_{j=1}^k \sum_{i=1}^k \Lambda_{\alpha_j}(\psi_i\varphi_j) = \sum_{j=1}^k \sum_{i=1}^k \Lambda_{\alpha_i}(\psi_i\varphi_j)$$
$$= \sum_{i=1}^n \Lambda_{\alpha_i}\left(\psi_i \sum_{j=1}^k \varphi_j\right) = 0.$$

It is easily seen that $\Lambda$ given by (2.3.1) is a linear form over $\mathcal{D}(\Omega)$. It remains to prove that $\Lambda$ is continuous. Indeed, let $K \subset \Omega$ be a compact set and $K \subset \bigcup_{j=1}^k \Omega_{\alpha_j}$. By virtue of Lemma 2.1.1 there exists $\psi \in \mathcal{D}\left(\bigcup_{j=1}^k \Omega_{\alpha_j}\right)$ such that $\psi(x) = 1$ for $x \in K$ and $0 \le f(x) \le 1$ if $x \in \bigcup_{j=1}^k \Omega_{\alpha_j}$. Taking into account Theorem 2.1.1 we can write $\psi = \sum_{j=1}^k \psi_j$, $\psi_j \in \mathcal{D}(\Omega_{\alpha_j})$. Finally we get

$$\Lambda(\varphi) = \Lambda\left(\sum_{j=1}^k \psi_j\varphi\right) = \sum_{j=1}^k \Lambda(\psi_j\varphi) = \sum_{j=1}^k \Lambda_{\alpha_j}(\psi_j\varphi). \tag{2.3.3}$$

From this, it follows immediately that if $\operatorname{supp} \varphi_\nu \subset K$ and the sequence $(\partial^\alpha \varphi_\nu)$ uniformly converges to zero for $\alpha \in \mathbb{N}^n$, then $\psi_j\varphi_\nu \to 0$ as $\nu \to \infty$ in $\mathcal{D}(\Omega_{\alpha_j})$. In view of (2.3.3) we infer that $\Lambda(\varphi_\nu) \to 0$ as $\nu \to \infty$. Thus, the proof of existence is complete. $\square$

EXAMPLE 2.3.1. Let $\Omega = \mathbb{R} - \{0\}$. Put $\Lambda(\varphi) = \int_\Omega e^{\frac{1}{x}} \varphi(x)\, dx$ for $\varphi \in \mathcal{D}(\Omega)$. We shall now show that there is no distribution $\widetilde{\Lambda}$ in $\mathcal{D}'$ such that $\widetilde{\Lambda}|_\Omega = \Lambda$. Indeed, let $\varphi \in \mathcal{D}(0,2)$, $\varphi \ge 0$ and $\int \varphi = 1$. Put $\varphi_\nu(x) = e^{-\frac{\nu}{2}} \nu\varphi(\nu x)$, $\operatorname{supp} \varphi_\nu \subset (0,2) \subset (-2,2)$. It is easy to check that $\varphi_\nu \to 0$ in $\mathcal{D}(-2,2)$. To obtain a contradiction suppose that there exists a distribution $\widetilde{\Lambda}$ in $\mathcal{D}'$ such that its restriction to $\Omega$ is equal to $\Lambda$, then we have

$$\widetilde{\Lambda}(\varphi_\nu) = \int_0^2 e^{(\frac{1}{x}-\frac{\nu}{2})}\nu\varphi(\nu x)\,dx.$$

An easy computation shows that

$$\widetilde{\Lambda}(\varphi_\nu) = \int_0^2 e^{\nu(\frac{1}{y}-\frac{1}{2})}\varphi(y)\,dy \geq \int_0^2 \varphi(y)dy = 1.$$

This contradicts the continuity of $\widetilde{\Lambda}$.

## 2.4. Support of distributions

Let $T \in \mathcal{D}'(\Omega)$. Let $\Omega'$ be a union of open sets $\Omega_\alpha \subset \Omega$, $\alpha \in \mathcal{A}$ such that $T$ vanishes in $\Omega_\alpha$. According to Theorem 2.3.1, $T$ vanishes in $\Omega'$.

This justifies taking the following

DEFINITION 2.4.1. labeld2.4.1 Let $\Lambda \in \mathcal{D}'(\Omega)$. Then the support of $\Lambda$ is the complement of the largest open set $\Omega' \subset \Omega$ with respect to $\Omega$ in which $\Lambda$ vanishes.

Note, that under this definition the support of a distribution is always closed. We shall usually take $\Omega = \mathbb{R}^n$. In this case, the support of a distribution may be whole $\mathbb{R}^n$ or it may be the empty set. For example, supp $\frac{1}{(\cdot)} = \mathbb{R}$ and supp $\delta = \{0\}$.

Some distributions are sensible for $\varphi$ belonging to larger test spaces than $\mathcal{D}(\Omega)$. For example $\delta(\varphi)$ is meaningful for $\varphi \in C(\mathbb{R}^n)$. We shall now show that an arbitrary distribution $\Lambda \in \mathcal{D}'(\Omega)$ may be extended on a subspace of $C^\infty(\Omega)$. Let

$$C_\Lambda^\infty(\Omega) := \{\varphi : \varphi \in C^\infty(\Omega),\ \mathrm{supp}\,\varphi \cap \mathrm{supp}\,\Lambda \Subset \Omega\}.$$

Note that $C_\Lambda^\infty(\Omega)$ is a vector subspace of $C^\infty(\Omega)$.

We are now in a position to prove the following

THEOREM 2.4.1. *If $\Lambda \in \mathcal{D}'(\Omega)$, then there exists exactly one linear form $\widetilde{\Lambda}$ : $C_\Lambda^\infty(\Omega) \to C$ such that*

(i) $\widetilde{\Lambda}(\varphi) = \Lambda(\varphi)$ *if $\varphi \in \mathcal{D}(\Omega)$,*

(ii) $\widetilde{\Lambda}(\varphi) = 0$ *if $\varphi \in C^\infty(\Omega)$ and $\mathrm{supp}\,\varphi \cap \mathrm{supp}\,\Lambda = \emptyset$.*

PROOF. Let $\mathrm{supp}\,\Lambda \cap \mathrm{supp}\,\varphi = K \Subset \Omega$. By virtue of Lemma 2.1.1 there exists a function $\psi \in \mathcal{D}(\Omega)$ such that $0 \leq \psi(x) \leq 1$ if $x \in \Omega$ and $\psi(x) = 1$ for $x \in \overline{B_\epsilon(K)}$, $B_\epsilon(K) \Subset \Omega$ for some $\epsilon > 0$ (if $K = \emptyset$ we take $\psi = 0$). Each function $\varphi \in C_\Lambda^\infty(\Omega)$ may be represented as follows $\varphi = \varphi_1 + \varphi_2$, where $\varphi_1 = \psi\varphi$ and $\varphi_2 = (1-\psi)\varphi$. Note that $\mathrm{supp}\,\Lambda \cap \mathrm{supp}\,\varphi_2 = \emptyset$. Since $\varphi \in C_\Lambda^\infty(\Omega)$, therefore $\varphi_1$ and $\varphi_2$ are in $C_\Lambda^\infty(\Omega)$, as well. Suppose now that there exists a linear form $\widetilde{\Lambda} : C_\Lambda^\infty(\Omega) \to C$ fulfilling (i) and (ii), then $\widetilde{\Lambda}(\varphi) = \widetilde{\Lambda}(\varphi_1) + \widetilde{\Lambda}(\varphi_2) = \Lambda(\psi\varphi)$. Put by definition

$$\widetilde{\Lambda}(\varphi) := \Lambda(\psi\varphi). \tag{2.4.1}$$

We have to show that $\Lambda(\psi\varphi)$ does not depend on $\psi$ if $\varphi \in C_\Lambda^\infty(\Omega)$. Indeed, assume that $\psi_1$ has the some properties as $\psi$, then $\widetilde{\Lambda}(\varphi) = \Lambda(\psi_1\varphi)$. Note that $\mathrm{supp}[(\psi_1 - \psi)\varphi] \cap \mathrm{supp}\,\Lambda = \emptyset$. Therefore $\Lambda(\psi_1\varphi) = \Lambda(\psi\varphi)$. One can prove that $\widetilde{\Lambda}$ is a linear form on $C_\Lambda^\infty(\Omega)$ satisfying (i) and (ii). $\square$

If $\operatorname{supp} \Lambda$ is compact, then one can take $K = \operatorname{supp} \Lambda$. Of course, $C_\Lambda^\infty(\Omega) = C^\infty(\Omega)$.

DEFINITION 2.4.2. A sequence $(\varphi_\nu)$, $\varphi_\nu \in C^\infty(\Omega)$ is said to be convergent in $C^\infty(\Omega)$ if the sequences $(\partial^\alpha \varphi_\nu)$ converge uniformly in every compact set $K \subset \Omega$ for each $\alpha \in \mathbb{N}^n$.

One can observe that if $\Lambda$ has a compact support then the linear form $\widetilde{\Lambda}$ : $C^\infty(\Omega) \to \mathbb{C}$ is continuous under the above defined convergence (see Theorem 1.1.1). Thus, we have shown that an arbitrary distribution $\Lambda \in \mathcal{D}'(\Omega)$ with a compact support may be extended by (2.4.1) on whole $C^\infty(\Omega)$ and the linear form $\widetilde{\Lambda}$ is continuous. Moreover, it may be extended on $C^\infty(\mathbb{R}^n)$ in the same way. Since $\widetilde{\Lambda}(\varphi) = \Lambda(\varphi)$ if $\varphi \in \mathcal{D}(\Omega)$, the distribution $\Lambda$ is also continuous in the above sense. Note that a sequence $(\varphi_\nu)$, $\varphi_\nu \in C^\infty(\Omega)$ converges to zero in $C^\infty(\Omega)$ iff for each compact set $K \subset \Omega$ and $m \in \mathbb{N}$, $\|\varphi_\nu\|_{m,K} \to 0$ as $\nu \to \infty$ (see (1.1.1)). Let

$$K_\nu = \left\{ x : x \in \Omega, \ |x| \leq \nu \text{ and } \operatorname{dist}(x, \partial\Omega) \geq \frac{1}{\nu} \right\},$$

then $\Omega = \bigcup_{\nu=1}^\infty K_\nu$. This indicates to us that the convergence in $C^\infty(\Omega)$ may be defined by means of the countable family $(\|\cdot\|_{m,K_m})$, $m = 1, 2, \ldots$ of the seminorms $\|\cdot\|_{m,K_m}$.

Now, we give a characterization of the linear continuous forms on $C^\infty(\Omega)$.

THEOREM 2.4.2. A linear form $\Lambda : C^\infty(\Omega) \to C$ is continuous with respect to $C^\infty(\Omega)$-convergence iff there exist a compact set $K \subset \Omega$, $m \in \mathbb{N}$ and $M > 0$ such that

$$|\Lambda(\varphi)| \leq M\|\varphi\|_{m,K} \qquad (2.4.2)$$

for $\varphi \in C^\infty(\Omega)$.

PROOF. The sufficiency is evident. We shall now prove the necessity. Assume that $\Lambda \in \mathcal{D}'(\Omega)$ and that there is not any compact set, $m \in \mathbb{N}$ and $M > 0$ such that (2.4.2) holds. Therefore for each $\nu \in \mathbb{N}$ there exists $\varphi_\nu \in C^\infty(\Omega)$ such that $|\Lambda(\varphi_\nu)| > \nu\|\varphi_\nu\|_{\nu,K_\nu}$. One can choose $\varphi_\nu$ so that $\Lambda(\varphi_\nu) = 1$. Thus, we have $\|\varphi_\nu\|_{\nu,K_\nu} < \frac{1}{\nu}$ for $\nu = 1, 2, \ldots$. Note that

$$\|\varphi_{\nu+\mu}\|_{\mu,K_\mu} \leq \|\varphi_{\nu+\mu}\|_{\nu+\mu,K_{\nu+\mu}} < \frac{1}{\nu+\mu} < \frac{1}{\nu}.$$

This implies that the sequence $(\varphi_\nu)$ converges to zero in $C^\infty(\Omega)$. This is a contradiction since $\Lambda$ is continuous. $\qquad \square$

We are able to give a characterization of distributions with compact supports.

THEOREM 2.4.3. A distribution $\Lambda \in \mathcal{D}'(\Omega)$ has a compact support iff (2.4.2) holds for some $K, m, M$ and $\varphi \in \mathcal{D}(\Omega)$.

PROOF. Taking into account the previous consideration, we only need to show the sufficiency. Indeed, suppose that $\operatorname{supp} \varphi \cap K = \emptyset$, then by (2.4.2), $\Lambda(\varphi) = 0$. Therefore $\operatorname{supp} \Lambda \subset K$. This proves the theorem. $\qquad \square$

The following theorem is important for applications.

THEOREM 2.4.4. $\mathcal{D}(\Omega)$ is dense in $C^\infty(\Omega)$.

PROOF. By Lemma 2.1.1 there exists a function $\psi_\nu \in \mathcal{D}(\Omega)$ such that $\psi_\nu(x) = 1$ for $x \in K_\nu$, $\nu = 1, 2, \ldots$. Take $f_\nu = \psi_\nu f$. Of course, $f_\nu \in \mathcal{D}(\Omega)$. Note that $\|f_\nu - f\|_{m, K_m} = 0$ for $\nu \geq m$. Therefore $f_\nu$ converges to $f$ in $C^\infty(\Omega)$.                    $\square$

COROLLARY 2.4.1. *If $\Lambda \in \mathcal{D}'(\Omega)$ has a compact support, then the linear form $\tilde{\Lambda} : C^\infty(\Omega) \to C$, defined by (2.4.1) may be obtained by means of the continuous extension $\Lambda$ from $\mathcal{D}(\Omega)$ onto $C^\infty(\Omega)$.*

## 2.5. Distributions of finite order

Let $\Lambda \in \mathcal{D}'(\Omega)$. In accordance with Theorem 1.1.1, for every compact set $K \subset \Omega$ there exist a nonnegative integer $m$ and $C > 0$ such that

$$|\Lambda(\varphi)| \leq C\|\varphi\|_{m, K} \tag{2.5.1}$$

if $\varphi \in \mathcal{D}(\Omega)$ and supp $\varphi \subset K$. For some distributions the index $m$ does not depend on $K$.

DEFINITION 2.5.1. A distribution $\Lambda$ is said to be of finite order if there exists a nonnegative integer $m$ (independent on $K$) such that (2.5.1) holds for $\varphi \in \mathcal{D}(\Omega)$, supp $\varphi \subset K$ (in general $C$ depends on $K$). The smallest integer $m \geq 0$ such that (2.5.1) holds is called the order of $\Lambda$.

EXAMPLE 2.5.1. Let $\Lambda : \mathcal{D}(\Omega) \to \mathbb{C}$ be a linear form fulfilling the following condition

$$\varphi \geq 0 \implies \Lambda(\varphi) \geq 0. \tag{2.5.2}$$

We shall now show that $\Lambda$ is a distribution of finite order. Indeed, let $K \subset \Omega$ be a compact set. In view of Lemma 2.1.1 there exists $\psi \in \mathcal{D}(\Omega)$, $0 \leq \psi(x) \leq 1$ and $\psi(x) = 1$ for $x \in K$. Let $\varphi \in \mathcal{D}(\Omega)$ and supp $\varphi \subset K$, then we have

$$-\|\varphi\|_{0, K} \psi(x) \leq \varphi(x) \leq \|\varphi\|_{0, K} \psi(x)$$

for $x \in \Omega$. Finally we get

$$|\Lambda(\varphi)| \leq C\|\varphi\|_{0, K},$$

where $C = \Lambda(\psi)$. Note that the Dirac measure $\delta$ satisfies (2.5.2). In this case, $C = 1$ for each compact set $K \subset \mathbb{R}^n$.

EXAMPLE 2.5.2. Put $\Lambda = \sum_{\nu \in \mathbb{Z}} D^\nu \delta_\nu$, where $\delta_\nu(\varphi) = \varphi(\nu)$. It is easily seen that the above series converges in $\mathcal{D}'(\mathbb{R})$. The distribution $\Lambda$ is not of finite order.

EXAMPLE 2.5.3. Every distribution with a compact support is of finite order (see Theorem 2.4.3).

We are in a position to prove the following

THEOREM 2.5.1. *Let $\Lambda$ be a distribution with the compact support $K$ of order $m$. If $\varphi \in C^\infty(\mathbb{R}^n)$ and its derivatives up to order $m$ vanish on $K$, then $\tilde{\Lambda}(\varphi) = 0$.*

PROOF. Take by definition

$$\eta(\epsilon) := \max_{|\alpha| \leq m} \sup_{x \in B_\epsilon(K)} |\partial^\alpha \varphi(x)|.$$

Note that if $\partial^\alpha \varphi$ vanishes on $K$, $|\alpha| \leq m$, then $\eta(\epsilon) \to 0$ as $\epsilon \to 0$. Let $\chi$ be the characteristic function of $B_{\frac{\epsilon}{2}}(K)$. Put $\chi_\epsilon = \chi * \psi_{\frac{\epsilon}{4}}$, where $\psi_\epsilon(x) = \epsilon^{-n}\psi\left(\frac{x}{\epsilon}\right)$, $\epsilon > 0$, $\psi \in \mathcal{D}(B_1(0))$, $\psi \geq 0$ and $\int \psi = 1$. It is easy to see that

$$\chi_\epsilon(x) = 1 \quad \text{if} \quad x \in B_{\frac{\epsilon}{4}}(K) \quad \text{and} \quad \chi_\epsilon(x) = 0 \quad \text{if} \quad x \notin \overline{B_{\frac{3\epsilon}{4}}(K)}.$$

An easy computation shows that

$$|\partial^\alpha \chi_\epsilon(x)| < \frac{M_\alpha}{\epsilon^{|\alpha|}} \quad \text{for} \quad x \in \mathbb{R}^n. \tag{2.5.3}$$

We shall now estimate $|\partial^\beta \varphi|$. Note that

$$\partial^\beta \varphi(x) = \int_{(e_1,x_0)}^{(e_1,x)} \partial^{\beta+e_1} \varphi(\tau, x_2, \ldots, x_n)\, d\tau, \tag{2.5.4}$$

where $x_0 \in \partial K$, $e_1 = (1,0,\ldots,0)$ and $(x,y) = (x_1 y_1, \ldots, x_n y_n)$. In accordance with the assumption we have

$$\partial^{\beta+e_1} \varphi(x) = 0 \quad \text{if} \quad x \in K$$

and

$$|\partial^{\beta+e_1} \varphi(x)| \leq \eta(\epsilon) \quad \text{if} \quad x \in B_\epsilon(K) - K \quad \text{when} \quad |\beta + e_1| \leq m.$$

Taking into account (2.5.4) we get

$$|\partial^\beta \varphi(x)| \leq \eta(\epsilon)\epsilon \quad \text{if} \quad x \in B_\epsilon(K) \quad \text{and} \quad |\beta + e_1| \leq m.$$

The same estimate is obtained if we integrate over $x_j$, $1 < j \leq m$, provided that $|\beta + e_j| \leq m$. Repeating this process with respect to the second variable we get

$$|\partial^\beta \varphi(x)| \leq \eta(\epsilon)\epsilon^2 \quad \text{if} \quad x \in B_\epsilon(K)$$

and $|\beta + e_1 + e_2| \leq m$. By iterating the process and integrating $m - |\beta|$ times, we obtain the estimate

$$|\partial^\beta \varphi(x)| \leq \eta(\epsilon)\epsilon^{m-|\beta|} \quad \text{for} \quad x \in B_\epsilon(K).$$

By combining this estimate with (2.5.3) and utilizing the Leibniz formula we get

$$|\partial^\alpha(\varphi\chi_\epsilon)(x)| \leq \eta(\epsilon) \sum_{0 \leq \beta \leq \alpha} \binom{\alpha}{\beta} \epsilon^{m-|\beta|} M_{\alpha-\beta} \epsilon^{-|\alpha-\beta|} = M\eta(\epsilon)\epsilon^{m-|\alpha|} \tag{2.5.5}$$

for $x \in \mathbb{R}^n$, where $M = \sum_{0 \leq \beta \leq \alpha} \binom{\alpha}{\beta} M_{\alpha-\beta}$. From this it follows

$$\|\chi_\epsilon \varphi\|_{m,A} \leq M\eta(\epsilon)$$

if $0 < \epsilon \leq 1$ and $A \supset B_\epsilon(K)$ is an arbitrary compact set. Since $\Lambda$ is of order $m$ with the compact support $K$, therefore for some $A \supset B_\epsilon(K)$ we have

$$|\Lambda(\varphi\chi_\epsilon)| \leq C\|\chi_\epsilon\varphi\|_{m,A} \leq CM\eta(\epsilon).$$

In accordance with the definition of $\widetilde{\Lambda}$ we get

$$|\widetilde{\Lambda}(\varphi)| \leq CM\eta(\epsilon).$$

Since $\eta(\epsilon) \to 0$ as $\epsilon \to 0$, therefore $\widetilde{\Lambda}(\varphi) = 0$. Thus, the proof is finished. $\quad\square$

As a corollary we obtain

THEOREM 2.5.2. *A distribution $\Lambda$ with the point support $\{x_0\}$ is a finite linear combination of Dirac's measure concentrated in $x_0$ and its derivatives.*

PROOF. Suppose that $\Lambda$ is of order $m$. Let $\varphi \in C^\infty$. The function $\varphi$ may be written as follows

$$\varphi(x) = \sum_{|\alpha| \leq m} \frac{\partial^\alpha \varphi(x_0)}{\alpha!} (x - x_0)^\alpha + R_m(x)$$

and $\partial^\alpha R_m(x_0) = 0$ if $|\alpha| \leq m$. In accordance with the above theorem we infer that

$$\Lambda(\varphi) = \sum_{|\alpha| \leq m} a_\alpha (-1)^{|\alpha|} \partial^\alpha \varphi(x_0)$$

for $\varphi \in \mathcal{D}$, where $a_\alpha = \frac{(-1)^{|\alpha|}}{\alpha!} \widetilde{\Lambda}((\cdot - x_0)^\alpha)$. Finally we have

$$\Lambda = \sum_{|\alpha| \leq m} a_\alpha D^\alpha \delta_{x_0}. \qquad \square$$

## 2.6. Cartesian products of Banach spaces

Let $E_j$, $j = 1, \ldots, n$ be a Banach space over $\mathbb{C}$. We recall that the Cartesian product $E = \mathrm{X}_{j=1}^n E_j$ of the vector spaces $E_j$ consists of all vectors $x = (x_1, \ldots, x_n)$, where $x_j \in E_j$. The element $x_j$ is called the $j$-th component of $x$. We introduce in the set $E$ addition and multiplication by scalar of elements belonging to $E$ as follows

$$x + y = (x_1 + y_1, \ldots, x_n + y_n)$$

if $x = (x_1, \ldots, x_n)$, $y = (y_1, \ldots, y_n)$,

$$\lambda x = (\lambda x_1, \ldots, \lambda x_n),$$

$\lambda \subset \mathbb{C}$. Obviously $E$ is a vector space under these operations. The map $\Pi_j$ from $E$ onto $E_j$ defined by $\Pi_j(x) = x_j$ is called the $j$-th projection. Take

$$\widetilde{E}_j = \{\tilde{x}_j : \tilde{x}_j = (0, \ldots, 0, x_j, 0, \ldots, 0), \ x_j \in E_j\}.$$

Note that $\Pi_j$ establishes an isomorphism from $\widetilde{E}_j$ onto $E_j$ ($\Pi_j(\tilde{x}_j) = x_j$). Every element $x \in E$ may be written in the form $x = \tilde{x}_1 + \cdots + \tilde{x}_n$. The representation of each element $x \in E$ in this form is unique. This means that the space $E$ is the direct sum of the subspaces $\widetilde{E}_j$ $j = 1, \ldots, n$. We shall write $E = \bigoplus_{j=1}^n \widetilde{E}_j$. In $E$ one can introduce two types of norms

(i) $\|x\|_p = \left( \sum_{j=1}^n \|x_j\|_j^p \right)^{\frac{1}{p}}$, $1 \leq p < \infty$,

(ii) $\|x\|_\infty = \max_{j=1,\ldots,n} \|x_j\|_j$, $p = \infty$,

where $\|x_j\|_j$ is the norm of $x_j$ in $E_j$. Note that

$$\|\tilde{x}_j\|_p = \|\Pi_j^{-1}(x_j)\|_p = \|x_j\|_j, \qquad 1 \leq p \leq \infty.$$

This means that the map $\Pi_j$ is a linear isometry from $\widetilde{E}_j$ onto $E_j$. Obviously, if $E_j$ ($j = 1, \ldots, n$) are the Banach spaces, then $E = \mathrm{X}_{j=1}^n E_j$ is also the Banach space under the norms $\|\cdot\|_p$, $1 \leq p \leq \infty$.

LEMMA 2.6.1. *Let $\Lambda$ be a linear continuous form on $E = \mathrm{X}_{j=1}^n E_j$ with respect to the norm $\|\cdot\|_p$, $1 \leq p \leq \infty$, then there exist continuous linear forms $\Lambda_j$ on $E_j$ such that*

$$\Lambda(x) = \sum_{j=1}^n \Lambda_j(x_j), \qquad (2.6.1)$$

*where $x = (x_1, \ldots, x_n)$, $x_j \in E_j$.*

PROOF. Since $E = \bigoplus_{j=1}^n \tilde{E}_j$, therefore $\Lambda(x) = \sum_{j=1}^n \Lambda(\tilde{x}_j)$. Note that $\Lambda(\tilde{x}_j) = \Lambda(\Pi_j^{-1}(x_j))$. Since $\Lambda$ is a continuous linear form on $E$, therefore the map $E_j \ni x_j \to (\Lambda \circ \Pi_j^{-1})(x_j)$ is a continuous linear form on $E_j$. We take $\Lambda_j = \Lambda \circ \Pi_j^{-1}$. Finally we obtain (2.6.1). $\qquad\square$

## 2.7. Some local representations of distributions

We start with presenting some information concerning complex measures.

DEFINITION 2.7.1. Let $\Omega$ be an open set in $\mathbb{R}^n$. We say that $\varphi \in C_0(\Omega)$ if it is continuous in $\Omega$, and for every $\epsilon > 0$ there exists a compact set $K \subset \Omega$ such that $|\varphi(x)| < \epsilon$ for $x \in \Omega - K$.

It is easily seen that $C_0(\Omega)$ is a vector space over the field of the complex numbers. Put

$$\|\varphi\|_{C_0(\Omega)} := \sup_{x \in \Omega} |\varphi(x)|.$$

An easy computation shows that $C_0(\Omega)$ is complete under the norm $\| \cdot \|_{C_0(\Omega)}$. We shall now recall that $C_c(\Omega)$ denotes the collection of all continuous complex functions on $\Omega$ whose support is compact (see Section 1.3).

We are able to prove the following

THEOREM 2.7.1. $C_c(\Omega)$ is dense in $C_0(\Omega)$.

PROOF. Given $f \in C_0(\Omega)$ and $\epsilon > 0$, then there exists a compact set $K \subset \Omega$ such that $|f(x)| < \epsilon$ outside $K$. Lemma 2.1.1 gives us a function $\psi \in C_c(\Omega)$ such that $0 \leq \psi(x) \leq 1$ for $x \in \Omega$ and $\psi(x) = 1$ on $K$. Put $g = f\psi$. Then $g \in C_c(\Omega)$ and $\|f - g\|_{C_0(\Omega)} < \epsilon$. Thus, the proof is finished. $\qquad\square$

Let $\mathcal{M}$ be a $\sigma$-algebra in $\Omega$ ([23]). A countable collection $\{E_\nu\}$ of the members of $\mathcal{M}$ is called a partition of $E$ if $E_i \cap E_j = \emptyset$ whenever $i \neq j$ and $E = \bigcup_{\nu=1}^\infty E_\nu$.

DEFINITION 2.7.2. A function $\mu : \mathcal{M} \to C$ such that

$$\mu(E) = \sum_{\nu=1}^\infty \mu(E_\nu) \tag{2.7.1}$$

for every partition $\{E_\nu\}$ of $E$ is called a complex measure on $\mathcal{M}$.

We shall now describe some general properties of the complex measures. Since the union of the sets $E_\nu$ is not changed if the subscripts $\nu$ are permuted, therefore the series (2.7.1) must be convergent absolutely. We are now in a position to define a finite positive measure $|\mu|$ on $\mathcal{M}$, putting

$$|\mu|(E) := \sup \sum_{\nu=1}^\infty |\mu(E_\nu)|, \tag{2.7.2}$$

the supremum being taken over all partitions $\{E_\nu\}$ of $E$.

DEFINITION 2.7.3. The measure $|\mu|$ is called the total variation of the measure $\mu$.

It is known that there exists a measurable function $h$ with respect to $\sigma$-algebra $\mathcal{M}$ such that $|h(x)| = 1$ for $x \in \Omega$ and $\mu(E) = \int_E h\,d|\mu|$ for $E \in \mathcal{M}$. It is therefore reasonable to define integration with respect to a complex measure $\mu$ by the formula

$$\int_E f\,d\mu = \int_E f h\,d|\mu|.$$

We shall now remember the Riesz representation theorem for linear continuous form on $C_0(\Omega)$, where $\Omega$ is an open set.

THEOREM 2.7.2. *To each linear continuous form $\Lambda$ on $C_0(\Omega)$ there corresponds a unique complex regular measure $\mu$ such that*

$$\Lambda(f) = \int_\Omega f\,d\mu = \int_\Omega f h\,d|\mu|$$

*for $f \in C_0(\Omega)$, where $|\mu|$ is a positive finite regular Borel measure and $h$ is a Borel function such that $|h(x)| = 1$ for $x \in \Omega$.*

PROOF. See [23].  □

THEOREM 2.7.3. *Let $\Omega' \Subset \Omega$ and $\Lambda \in \mathcal{D}'(\Omega)$, then there exist $m \in \mathbb{N}$ and complex regular Borel measures $\mu_\alpha$ on $\Omega'$ such that*

$$\Lambda(\varphi) = \sum_{|\alpha| \leq m} \int_{\Omega'} (-1)^{|\alpha|} \partial^\alpha \varphi \, d\mu_\alpha,$$

*if $\operatorname{supp} \varphi \subset \overline{\Omega'}$.*

PROOF. Let $\Omega' \Subset \Omega$ and $\Lambda \in \mathcal{D}'(\Omega)$, then according to Theorem 1.1.1, $m \in \mathbb{N}$ and $C > 0$ exist such that

$$|\Lambda(\varphi)| \leq C\|\varphi\|_{m,\overline{\Omega'}}$$

if $\operatorname{supp} \varphi \subset \overline{\Omega'}$. Consider the Cartesian product

$$C_0^N(\Omega') = \underset{|\alpha| \leq m}{\text{X}}\, C_0(\Omega'), \quad N = \sum_{|\alpha| \leq m} 1.$$

The elements of $C_0^N(\Omega')$ will be denoted by $(\varphi_\alpha)$. The component of $(\varphi_\alpha)$ corresponding to subscript $\alpha$ will be denoted by $\varphi_\alpha$. Let us endow the Cartesian product $C_0^N(\Omega')$ with the norm

$$\|(\varphi_\alpha)\|_{C_0^N(\Omega')} := \|(\varphi_\alpha)\|_\infty$$

(see (ii) in Section 2.6). According to the previous considerations we infer that $C_0^N(\Omega')$ is a Banach space with respect to this norm. We shall now define a map $\mathcal{K}$ from $\mathcal{D}(\Omega')$ into $C_0^N(\Omega')$ as follows

$$\varphi \to \mathcal{K}(\varphi) = \left( (-1)^{|\alpha|} \partial^\alpha \varphi \right).$$

If we endow the set $\mathcal{D}(\Omega')$ with the norm $\|\cdot\|_{m,\overline{\Omega'}}$, then the map $\mathcal{K}$ is an isomorphism from $\mathcal{D}(\Omega')$ onto $\mathcal{K}(\mathcal{D}(\Omega')) \subset C_0^N(\Omega')$, moreover

$$\|\varphi\|_{m,\overline{\Omega'}} = \| \left( (-1)^{|\alpha|} \partial^\alpha \varphi \right) \|_{C_0^N(\Omega')}.$$

Consider a new functional $\widetilde{\Lambda} : \mathcal{K}(\mathcal{D}(\Omega')) \to \mathbb{C}$ defined by means of the formula

$$\widetilde{\Lambda} \left( ((-1)^{|\alpha|} \partial^\alpha \varphi) \right) := \Lambda(\varphi).$$

The definition of $\widetilde{\Lambda}$ is meaningful, because the map $\mathcal{K}$ is an injection of $\mathcal{D}(\Omega')$ into $C_0^N(\Omega')$. Since $\mathcal{K}$ is an isometry isomorphism from $\mathcal{D}(\Omega')$ onto $\mathcal{K}(\mathcal{D}(\Omega'))$, therefore $\widetilde{\Lambda}$ is a continuous linear form on $\mathcal{K}(\mathcal{D}(\Omega'))$. By the Hahn–Banach theorem the linear form $\widetilde{\Lambda}$ may be extended from $\mathcal{K}(\mathcal{D}(\Omega'))$ on $C_0^N(\Omega')$ so that

$$|\widetilde{\Lambda}((\varphi_\alpha))| \leq C\|(\varphi_\alpha)\|_{C_0^N(\Omega')}$$

for $(\varphi_\alpha) \in C_0^N(\Omega')$. By virtue of Lemma 2.6.1 and Theorem 2.7.2 we obtain

$$\widetilde{\Lambda}((\varphi_\alpha)) = \sum_{|\alpha| \leq m} \int_{\Omega'} \varphi_\alpha \, d\mu_\alpha,$$

where $\mu_\alpha$ are complex regular Borel measures on $\Omega'$. Hence, for $((-1)^{|\alpha|}\partial^\alpha\varphi) \in \mathcal{K}(\mathcal{D}(\Omega'))$ we have $\widetilde{\Lambda}((-1)^{|\alpha|}\partial^\alpha\varphi) = \Lambda(\varphi)$. Thus, finally we obtain

$$\Lambda(\varphi) = \sum_{|\alpha| \leq m} \int_{\Omega'} (-1)^{|\alpha|}\partial^\alpha\varphi \, d\mu_\alpha.$$

This statement finishes the proof.                                              $\square$

CHAPTER 3

# TENSOR PRODUCTS AND CONVOLUTION PRODUCTS

### 3.1. Regularization of distributions

In this chapter we shall assume, for simplicity of consideration, that the distributions are defined on whole $\mathbb{R}^n$. Let $\varphi$ be in $\mathcal{D}(\mathbb{R}^m \times \mathbb{R}^n)$ and $\Lambda \in \mathcal{D}'(\mathbb{R}^m)$. Let us consider the following map

$$\mathbb{R}^n \ni y \to \Lambda(\varphi(\cdot, y)).$$

Put by definition

$$\psi(y) := \Lambda(\varphi(\cdot, y)).$$

THEOREM 3.1.1. *If $\Lambda \in \mathcal{D}'(\mathbb{R}^m)$ and $\varphi \in \mathcal{D}(\mathbb{R}^m \times \mathbb{R}^n)$, then $\psi$ is in $\mathcal{D}(\mathbb{R}^n)$.*

PROOF. Let $\operatorname{supp}\varphi \subset I_m \times I_n$, where $I_m$ and $I_n$ are compact intervals in $\mathbb{R}^m$ and $\mathbb{R}^n$, respectively. Since the function $\partial_x^\alpha \varphi(\cdot, \cdot)$ is uniformly continuous on $\mathbb{R}^m \times \mathbb{R}^n$, therefore

$$\left| \frac{\partial^{|\alpha|}}{\partial x^\alpha} \varphi(x, y+h) - \frac{\partial^{|\alpha|}}{\partial x^\alpha} \varphi(x, y) \right|, \quad \alpha \in \mathbb{N}^m$$

converges uniformly to zero as $h \to 0$ with respect to $x \in I_m$ for fixed $y$. From this, it follows that $\Lambda(\varphi(\cdot, y+h))$ tends to $\Lambda(\varphi(\cdot, y))$ as $h \to 0$. It is easily seen that $\operatorname{supp}\psi \subset I_n$. It remains to prove that $\psi$ is smooth. Taking into account the above consideration one can remark that $|h^{-1}(\varphi(\cdot, y+he_i) - \varphi(\cdot, y)) - \partial_y^{e_i}\varphi(\cdot, y)|$ tends to zero as $h \to 0$ in $\mathcal{D}(\mathbb{R}^m)$ for fixed $y$. Therefore

$$\Lambda\left( \frac{1}{h}(\varphi(\cdot, y+he_i) - \varphi(\cdot, y)) \right) \to \Lambda(\partial_y^{e_i}\varphi(\cdot, y))$$

as $h \to 0$. This gives us

$$\frac{\partial}{\partial y_i}\psi(y) = \Lambda\left( \frac{\partial}{\partial y_i}\varphi(\cdot, y) \right).$$

The continuity of $\partial^{e_i}\psi$ is obvious. By induction we have

$$\frac{\partial^{|\beta|}}{\partial y^\beta}\psi(y) = \Lambda\left( \frac{\partial^{|\beta|}}{\partial y^\beta}\varphi(\cdot, y) \right).$$

Thus, the proof is finished. □

Let $\varphi$ be in $\mathcal{D}$ and $\Lambda$ be in $\mathcal{D}'$. Put by definition

$$\Lambda_\varphi(x) := \Lambda(\varphi(x - \cdot)) \tag{3.1.1}$$

for $x \in \mathbb{R}^n$. We shall now present the following theorem which is parallel with Theorem 3.1.1.

THEOREM 3.1.2. *If $\Lambda \in \mathcal{D}'$ and $\varphi \in \mathcal{D}$ then $\Lambda_\varphi$ is a smooth function and*

$$\frac{\partial^{|\alpha|}}{\partial x^\alpha} \Lambda_\varphi(x) = \Lambda\left(\frac{\partial^{|\alpha|}}{\partial x^\alpha}\varphi(x - \cdot)\right).$$

Since the proof of this theorem is the same as the proof of Theorem 3.1.1 we shall not give it here.

DEFINITION 3.1.1. *The function $\Lambda_\varphi$ given by (3.1.1) will be called the regularization of $\Lambda$.*

Of course, if $\Lambda$ has a compact support then $\Lambda_\varphi$ is in $\mathcal{D}$ for each $\varphi$ in $\mathcal{D}$.

EXAMPLE 3.1.1. For $\Lambda = \delta$, where $\delta$ is the Dirac measure concentrated at zero we have $\delta_\varphi(x) = \varphi(x)$.

EXAMPLE 3.1.2. If $\Lambda = f$ and $f \in L^1_{loc}(\mathbb{R}^n)$, then

$$\Lambda_\varphi(x) = \int_{\mathbb{R}^n} f(y)\varphi(x - y)dy = (f * \varphi)(x).$$

This justifies the following notation $\Lambda * \varphi := \Lambda_\varphi$.

The following theorem is now an easy consequence of the preceding theorem.

THEOREM 3.1.3. *If $\Lambda$ is in $\mathcal{D}'$ and $\varphi$ belongs to $\mathcal{D}$, then*

$$(D^\alpha \Lambda * \varphi)(x) = (\Lambda * \partial^\alpha \varphi)(x) = \partial^\alpha(\Lambda * \varphi)(x). \tag{3.1.2}$$

PROOF. In accordance with Theorem 3.1.2 we have

$$\partial^\alpha(\Lambda * \varphi)(x) = \Lambda(\partial^\alpha \varphi(x - \cdot)) = (\Lambda * \partial^\alpha \varphi)(x). \tag{3.1.3}$$

Note that

$$(D^\alpha \Lambda * \varphi)(x) = D^\alpha \Lambda(\varphi(x - \cdot)) = (-1)^{|\alpha|}\Lambda_y\left(\frac{\partial^{|\alpha|}}{\partial y^\alpha}\varphi(x - y)\right)$$

$$= \Lambda_y\left(\frac{\partial^{|\alpha|}}{\partial x^\alpha}\varphi(x - y)\right) = \Lambda(\partial^\alpha \varphi(x - \cdot)). \tag{3.1.4}$$

In view of (3.1.3) and (3.1.4) we have (3.1.2). □

The next theorem is very important for applications.

THEOREM 3.1.4. *If $\varphi$ and $\psi$ belong to $\mathcal{D}$ but $\Lambda$ is in $\mathcal{D}'$, then*

$$(\Lambda * \varphi) * \psi = \Lambda * (\varphi * \psi).$$

PROOF. Put

$$\sigma_h(t) := \sum_{\nu \in \mathbb{Z}^n_h} \varphi(t - h\nu)\psi(h\nu)h^n, \tag{3.1.5}$$

where $\nu = (\nu_1,\ldots,\nu_n)$, $\nu_i \in \mathbb{Z}$, $h > 0$ and $\mathbb{Z}_h^n = \{\nu : h\nu \in \operatorname{supp}\psi\}$. Note that $\sigma_h(t)$ is a Riemann sum of the convolution $\varphi * \psi$ at point $t$. Differentiating $\alpha$-times equality (3.1.5) we obtain

$$\partial^\alpha \sigma_h(t) = \sum_{\nu \in \mathbb{Z}_h^n} \partial^\alpha \varphi(t - h\nu)\psi(h\nu)h^n. \tag{3.1.6}$$

By virtue of the mean value theorem we get

$$\partial^\alpha(\varphi * \psi)(t) = \int_{\mathbb{R}^n} \frac{\partial^{|\alpha|}}{\partial t^\alpha}\varphi(t - \tau)\psi(\tau)\,d\tau = \sum_{\nu \in \mathbb{Z}_h^n} \frac{\partial^{|\alpha|}}{\partial t^\alpha}\varphi(t - \Theta_\nu h\nu)\psi(\Theta_\nu h\nu)h^n, \tag{3.1.7}$$

where $0 \le \Theta_\nu \le 1$. By (3.1.6) and (3.1.7) we get

$$\left|\frac{\partial^{|\alpha|}}{\partial t^\alpha}\sigma_h(t) - \frac{\partial^{|\alpha|}}{\partial t^\alpha}(\varphi * \psi)(t)\right|$$
$$\le \sum_{\nu \in \mathbb{Z}_h^n} \left|\frac{\partial^{|\alpha|}}{\partial t^\alpha}\varphi(t - h\nu)\psi(h\nu) - \frac{\partial^{|\alpha|}}{\partial t^\alpha}\varphi(t - \Theta_\nu h\nu)\psi(\Theta_\nu h\nu)\right| h^n.$$

Put by definition $g(t,\tau) := \varphi(t - \tau)\psi(\tau)$. It is easy to check that

$$\operatorname{supp} g \subset \overline{(\operatorname{supp}\varphi + \operatorname{supp}\psi)} \times \operatorname{supp}\psi.$$

Therefore the function $g$ is uniformly continuous in $\mathbb{R}^n \times \mathbb{R}^n$. Moreover, there exists $M > 0$ such that $\sum_{\nu \in \mathbb{Z}_h^n} h^n < M$ for $h > 0$. From this, it follows that

$$\left|\frac{\partial^{|\alpha|}}{\partial t^\alpha}\sigma_h(t) - \frac{\partial^{|\alpha|}}{\partial t^\alpha}(\varphi * \psi)(t)\right| < \epsilon \tag{3.1.8}$$

for small $h$ and $t \in \mathbb{R}^n$. It means that $\frac{\partial^{|\alpha|}}{\partial t^\alpha}\sigma_h(t)$ tends uniformly to $\frac{\partial^{|\alpha|}}{\partial t^\alpha}(\varphi * \psi)(t)$ as $h \to 0$ in $\mathbb{R}^n$. Of course, $\frac{\partial^{|\alpha|}}{\partial t^\alpha}\sigma_h(x - t)$ also tends uniformly to $\frac{\partial^{|\alpha|}}{\partial t^\alpha}(\varphi * \psi)(x - t)$ as $h \to 0$ with respect to $t \in \mathbb{R}^n$ and fixed $x$. It is easily seen that

$$\operatorname{supp}\sigma_h(x - \cdot) \subset x - \overline{(\operatorname{supp}\varphi + \operatorname{supp}\psi)}.$$

Taking into account these statements we infer that

$$\Lambda(\sigma_h(x - \cdot)) \to \Lambda((\varphi * \psi)(x - \cdot)) = (\Lambda * (\varphi * \psi))(x)$$

as $h \to 0$. Of course,

$$\Lambda(\sigma_h(x - \cdot)) = \sum_{\nu \in \mathbb{Z}_h^n} \Lambda((\varphi(x - \cdot - h\nu))\psi(h\nu)h^n.$$

Note that the expression on the right side of this equality is a Riemann sum of the integral

$$\int_{\mathbb{R}^n} (\Lambda * \varphi)(x - \tau)\psi(\tau)d\tau.$$

Therefore $\Lambda(\sigma_h(x - \cdot))$ tends to $((\Lambda * \varphi) * \psi)(x)$ as $h \to 0$. This completes the proof. $\square$

We now conclude our consideration with an interesting application of the preceding theorem. As usually, let $\psi \in \mathcal{D}$, $\operatorname{supp}\psi \subset B_1(0)$ and $\int \psi = 1$. Put $\psi_\epsilon(x) := \epsilon^{-n}\psi\left(\frac{x}{\epsilon}\right)$.

THEOREM 3.1.5. *If $\Lambda \in \mathcal{D}'$, then $\Lambda * \psi_\epsilon$ tends to $\Lambda$ as $\epsilon \to 0$ in $\mathcal{D}'$.*

PROOF. We have to show that

$$\lim_{\epsilon \to 0} \int_{\mathbb{R}^n} (\Lambda * \psi_\epsilon)(x)\varphi(x)\, dx = \Lambda(\varphi)$$

for $\varphi \in \mathcal{D}$. Indeed, if we put $\check{\varphi}(\cdot) := \varphi(-\cdot)$, then

$$\int_{\mathbb{R}^n} (\Lambda * \psi_\epsilon)(x)\varphi(x)\, dx = \int_{\mathbb{R}^n} (\Lambda * \psi_\epsilon)(x)\check{\varphi}(0 - x)\, dx$$
$$= ((\Lambda * \psi_\epsilon) * \check{\varphi})(0).$$

By Theorem 3.1.4 we have

$$\int_{\mathbb{R}^n} (\Lambda * \psi_\epsilon)(x)\varphi(x)\, dx = \Lambda((\psi_\epsilon * \check{\varphi})(0 - \cdot)).$$

Because $\partial^\alpha(\varphi * \psi) = \partial^\alpha\varphi * \psi$, taking into account Theorem 1.3.2, we infer that $\psi_\epsilon * \check{\varphi}$ tends to $\check{\varphi}$ as $\epsilon \to 0$ in $\mathcal{D}$. Therefore $\Lambda(\psi_\epsilon * \check{\varphi}) \to \Lambda(\check{\varphi}(-\cdot)) = \Lambda(\varphi)$ as $\epsilon \to 0$. Finally we have

$$\int_{\mathbb{R}^n} (\Lambda * \psi_\epsilon)(x)\varphi(x)\, dx \to \Lambda(\varphi)$$

as $\epsilon \to 0$.                                                                    $\square$

This theorem tells us that the space $C^\infty(\mathbb{R}^n)$ is dense in $\mathcal{D}'(\mathbb{R}^n)$.

## 3.2. A characterization of convolution operators

Let $\Lambda$ be in $\mathcal{D}'$ and $\varphi$ be in $\mathcal{D}$. We have proved that $\Lambda * \varphi$ is infinitely differentiable in $\mathbb{R}^n$. Put by definition

$$U_\Lambda(\varphi) := \Lambda * \varphi.$$

THEOREM 3.2.1. *The map $U_\Lambda : \mathcal{D} \to C^\infty$ is a continuous linear operator and*

$$\Lambda * (\tau_h\varphi) = \tau_h(\Lambda * \varphi) \tag{3.2.1}$$

*for every $h \in \mathbb{R}^n$.*

PROOF. Of course, $U_\Lambda$ is a linear operator from $\mathcal{D}$ into $C^\infty$. Note that

$$(\Lambda * (\tau_h\varphi))(x) = (\Lambda * \varphi(\cdot - h))(x) = \Lambda(\varphi(x - \cdot - h))$$
$$= \tau_h(\Lambda(\varphi(x - \cdot))) = \tau_h(\Lambda * \varphi)(x).$$

Therefore, we only need to prove that $U_\Lambda$ is continuous. Let $\varphi$ be in $\mathcal{D}$. Assume that $\operatorname{supp}\varphi \subset B_r(0)$. Fix a closed ball $B_R(0)$. Obviously

$$B_r(0) + B_R(0) \subset \overline{B_{r+R}(0)}.$$

Since $\Lambda$ is in $\mathcal{D}'$, therefore there exist a nonnegative integer $m$ and $M > 0$ such that

$$|\partial^\alpha(\Lambda * \varphi)(x)| = |\Lambda(\partial^\alpha\varphi(x - \cdot))| \leq M\|\partial^\alpha\varphi\|_{m,\overline{B_{R+r}}}$$

for $x \in B_R(0)$. This implies that $U_\Lambda$ is continuous. Thus, the proof is finished. $\square$

We shall now show that this theorem admits a converse theorem.

THEOREM 3.2.2. *If $U : \mathcal{D} \to C^\infty$ is a continuous linear operator and $U\tau_x = \tau_x U$ for $x \in \mathbb{R}^n$, then there exists a distribution $\Lambda$ such that $U\varphi = \Lambda * \varphi$ for $\varphi \in \mathcal{D}$.*

PROOF. Define a new map as follows

$$\varphi \to (U\check{\varphi})(0),$$

where $\check{\varphi}(x) = \varphi(-x)$. It is easily seen that this map is a distribution. Put $\Lambda(\varphi) := (U\check{\varphi})(0)$. It remains to prove that $(U\varphi)(x) = (\Lambda * \varphi)(x)$. Indeed

$$(U\varphi)(-h) = \tau_h(U\varphi)(0) = U(\tau_h\varphi)(0) = \Lambda(\check{\tau_h\varphi}) = \Lambda(\check{\varphi}(\cdot - h))$$
$$= \Lambda(\varphi(-\cdot - h)) = (\Lambda * \varphi)(-h).$$

Put $x = -h$, then $(U\varphi)(x) = (\Lambda * \varphi)(x)$. This completes the proof. $\square$

## 3.3. Tensor product of distributions

Let $f$ and $g$ be in $C(\mathbb{R}^m)$ and $C(\mathbb{R}^n)$, respectively. Put by definition

$$(f \otimes g)(x, y) := f(x)g(y)$$

for $x \in \mathbb{R}^m$ and $y \in \mathbb{R}^n$. The function $f \otimes g$ is called the tensor product of $f$ and $g$. For arbitrary functions $\varphi_1$ and $\varphi_2$ belonging to $\mathcal{D}(\mathbb{R}^m)$ and $\mathcal{D}(\mathbb{R}^n)$, respectively we have

$$\int_{\mathbb{R}^m \times \mathbb{R}^n} (f \otimes g)(x, y)\varphi_1(x)\varphi_2(y)\,dx\,dy = \int_{\mathbb{R}^m} f(x)\varphi_1(x)dx \int_{\mathbb{R}^n} g(y)\varphi_2(y)\,dy. \quad (3.3.1)$$

Equality (3.3.1) defines a distribution in $\mathbb{R}^m \times \mathbb{R}^n$. This observation suggests to us the following

THEOREM 3.3.1. *If $S \in \mathcal{D}'(\mathbb{R}^m)$ and $T \in \mathcal{D}'(\mathbb{R}^n)$, then there exists exactly one distribution $K$ in $\mathcal{D}'(\mathbb{R}^{m+n})$ such that*

$$K(\varphi_1 \otimes \varphi_2) = S(\varphi_1)T(\varphi_2) \quad (3.3.2)$$

*for $\varphi_1 \in \mathcal{D}(\mathbb{R}^m)$ and $\varphi_2 \in \mathcal{D}(\mathbb{R}^n)$. Moreover,*

$$K(\varphi(\cdot, \cdot)) = S_x(T_y(\varphi(x, y))) = T_y(S_x(\varphi(x, y))) \quad (3.3.3)$$

*for $\varphi \in \mathcal{D}(\mathbb{R}^m \times \mathbb{R}^n)$.*

PROOF. We start with the proof of uniqueness. Note that we only need to show that if

$$K(\varphi_1 \otimes \varphi_2) = S(\varphi_1)T(\varphi_2) = 0$$

for $\varphi_1 \in \mathcal{D}(\mathbb{R}^m)$ and $\varphi_2 \in \mathcal{D}(\mathbb{R}^n)$, then $K(\varphi) = 0$ for all $\varphi \in \mathcal{D}(\mathbb{R}^m \times \mathbb{R}^n)$. Let $\psi_1$ and $\psi_2$ be in $\mathcal{D}(\mathbb{R}^m)$ and $\mathcal{D}(\mathbb{R}^n)$, respectively and $\int_{\mathbb{R}^m} \psi_1 = \int_{\mathbb{R}^n} \psi_2 = 1$. Put by definition

$$\psi_\epsilon(x,y) := \epsilon^{-(m+n)} \psi_1 \left(\frac{x}{\epsilon}\right) \psi_2 \left(\frac{y}{\epsilon}\right).$$

Further, suppose that $K$ belongs to $\mathcal{D}'(\mathbb{R}^{m+n})$ and (3.3.2) holds. Of course, $\epsilon^{-m} \psi_1 \left(\frac{x-\cdot}{\epsilon}\right) \in \mathcal{D}(\mathbb{R}^m)$ for each $x \in \mathbb{R}^m$ and $\epsilon^{-n} \psi_2 \left(\frac{y-\cdot}{\epsilon}\right) \in \mathcal{D}(\mathbb{R}^n)$ for each $y \in \mathbb{R}^n$. Because of the assumption, note that

$$(K * \psi_\epsilon)(x,y) = K \left(\epsilon^{-m} \psi_1 \left(\frac{x-\cdot}{\epsilon}\right) \epsilon^{-n} \psi_2 \left(\frac{y-\cdot}{\epsilon}\right)\right)$$

$$= S \left(\epsilon^{-m} \psi_1 \left(\frac{x-\cdot}{\epsilon}\right)\right) T \left(\epsilon^{-n} \psi_2 \left(\frac{y-\cdot}{\epsilon}\right)\right) = 0$$

for $x \in \mathbb{R}^m$ and $y \in \mathbb{R}^n$. In view of Theorem 3.1.5

$$0 = \int_{\mathbb{R}^m \times \mathbb{R}^n} (K * \psi_\epsilon)(x,y)\varphi(x,y)dxdy \to K(\varphi(\cdot,\cdot))$$

as $\epsilon \to 0$. This statement finishes the proof of uniqueness.

We are left with the task of determining $K$. In order to do it assume that $\varphi(\cdot,\cdot)$ is in $\mathcal{D}(\mathbb{R}^m \times \mathbb{R}^n)$ and $\operatorname{supp} \varphi(\cdot,\cdot) \subset I_m \times I_n$ (see Section 3.1). Put by definition

$$\psi_1(x) := T(\varphi(x,\cdot))$$

and

$$\psi_2(y) := S(\varphi(\cdot,y)).$$

Taking into account the proof of Theorem 3.1.1 one can observe that $\psi_1$ is in $\mathcal{D}(I_m)$ and $\psi_2$ is in $\mathcal{D}(I_n)$. Moreover,

$$\frac{\partial^{|\alpha|}}{\partial x^\alpha} \psi_1(x) = T \left(\frac{\partial^{|\alpha|}}{\partial x^\alpha} \varphi(x,\cdot)\right) \quad \text{and} \quad \frac{\partial^{|\beta|}}{\partial y^\beta} \psi_2(y) = S \left(\frac{\partial^{|\beta|}}{\partial y^\beta} \varphi(\cdot,y)\right).$$

Since $S \in \mathcal{D}'(\mathbb{R}^m)$ and $T \in \mathcal{D}'(\mathbb{R}^n)$, therefore there exist non-negative integers $m_1$ and $m_2$ as well as $C_1$ and $C_2$ such that

$$|S(\psi_1)| \leq C_1 \|\psi_1\|_{m_1,I_m} \tag{3.3.4}$$

for $\varphi_1 \in \mathcal{D}(\mathbb{R}^m)$ if $\operatorname{supp} \psi_1 \subset I_m$ and

$$|T(\psi_2)| \leq C_2 \|\psi_2\|_{m_2,I_n} \tag{3.3.5}$$

for $\varphi_2 \in \mathcal{D}(\mathbb{R}^n)$ if $\operatorname{supp} \psi_2 \subset I_n$. Finally we have

$$|S_x(T_y(\varphi(x,y)))| \leq C_1 C_2 \|\varphi(\cdot,\cdot)\|_{m_1+m_2,I_m \times I_n} \tag{3.3.6}$$

and

$$|T_y(S_x(\varphi(x,y)))| \leq C_1 C_2 \|\varphi(\cdot,\cdot)\|_{m_1+m_2,I_m \times I_n} \tag{3.3.7}$$

for $\varphi \in \mathcal{D}(\mathbb{R}^m \times \mathbb{R}^n)$ if $\operatorname{supp} \varphi(\cdot,\cdot) \subset I_m \times I_n$. Of course, the maps

$$\varphi \to S_x(T_y(\varphi(x,y))) \quad \text{and} \quad \varphi \to T_y(S_x(\varphi(x,y))) \tag{3.3.8}$$

are linear forms on $\mathcal{D}(\mathbb{R}^m \times \mathbb{R}^n)$. By virtue of (3.3.6) and (3.3.7) we infer that the forms (3.3.8) are continuous on $\mathcal{D}(\mathbb{R}^m \times \mathbb{R}^n)$. Let

$$K_1(\varphi(\cdot,\cdot)) := S_x(T_y(\varphi(x,y)))$$

and

$$K_2(\varphi(\cdot,\cdot)) := T_y(S_x(\varphi(x,y)))$$

for $\varphi \in \mathcal{D}(\mathbb{R}^m \times \mathbb{R}^n)$. Note that

$$K_1(\varphi_1 \otimes \varphi_2) = K_2(\varphi_1 \otimes \varphi_2).$$

Analysis similar to that in the proof of uniqueness shows that for an arbitrary $\varphi \in \mathcal{D}(\mathbb{R}^m \times \mathbb{R}^n)$, $(K_1 - K_2)(\varphi(\cdot,\cdot)) = 0$. Put

$$K := K_1 = K_2.$$

Thus, we have proved (3.3.2) and (3.3.3). $\qquad\square$

After the above considerations we shall now give a definition of the tensor product of two distributions.

DEFINITION 3.3.1. Let $S$ and $T$ be in $\mathcal{D}'(\mathbb{R}^n)$ and $\mathcal{D}'(\mathbb{R}^m)$, respectively. Then the linear form $K$ defined above belongs to $\mathcal{D}'(\mathbb{R}^m \times \mathbb{R}^n)$ and fulfils both (3.3.2) and (3.3.3). The distribution $K$ is called the tensor product of $S$ and $T$.

The tensor product of $S$ and $T$ is usually denoted by $S \otimes T$. From (3.3.2) and (3.3.3) we obtain immediately the following theorems:

THEOREM 3.3.2. If $S \in \mathcal{D}'(\mathbb{R}^m)$ and $T \in \mathcal{D}'(\mathbb{R}^n)$, then

$$S \otimes T = T \otimes S \in \mathcal{D}'(\mathbb{R}^{m+n}).$$

THEOREM 3.3.3. If $Q \in \mathcal{D}'(\mathbb{R}^{n_1})$, $S \in \mathcal{D}'(\mathbb{R}^{n_2})$ and $T \in \mathcal{D}'(\mathbb{R}^{n_3})$, then

$$(Q \otimes S) \otimes T = Q \otimes (S \otimes T) \in \mathcal{D}'(\mathbb{R}^{n_1+n_2+n_3}).$$

## 3.4. Differentiation and support of tensor product

Let $\alpha = (\alpha_1, ..., \alpha_m)$ and $\beta = (\beta_1, ..., \beta_n)$. Put $(\alpha, \beta) = (\alpha_1, ..., \alpha_m, \beta_1, ..., \beta_n)$. We know that if $f \in C^\infty(\mathbb{R}^m)$ and $g \in C^\infty(\mathbb{R}^n)$, then

$$\frac{\partial^{|\alpha|}}{\partial x^\alpha}\left(\frac{\partial^{|\beta|}}{\partial y^\beta}(f \otimes g)(x,y)\right) = \frac{\partial^{|\alpha|}}{\partial x^\alpha}f(x)\frac{\partial^{|\beta|}}{\partial y^\beta}g(y)$$

for every $\alpha \in \mathbb{N}^m$ and $\beta \in \mathbb{N}^n$. We shall now show that the same formula holds for the tensor product $S \otimes T$ of the arbitrary distributions $S$ and $T$.

THEOREM 3.4.1. If $S \in \mathcal{D}'(\mathbb{R}^m)$ and $T \in \mathcal{D}'(\mathbb{R}^n)$, then

$$D_x^\alpha D_y^\beta(S_x \otimes T_y)(\varphi(x,y)) = (D_x^\alpha S_x \otimes D_y^\beta T_y)(\varphi(x,y)). \tag{3.4.1}$$

PROOF. Indeed, we have

$$D_x^\alpha D_y^\beta (S_x \otimes T_y)\varphi(x,y) = D^\alpha \left( (-1)^{|\beta|} (S_x \otimes T_y) \left( \frac{\partial^{|\beta|}}{\partial y^\beta} \varphi(x,y) \right) \right)$$

$$= D^\alpha \left( (-1)^{|\beta|} \left( S_x \left( T_y \left( \frac{\partial^{|\beta|}}{\partial y^\beta} \varphi(x,y) \right) \right) \right) \right)$$

$$= D^\alpha S_x (D^\beta T_y(\varphi(x,y)))$$

$$= (-1)^{|\alpha|} S_x \left( \frac{\partial^{|\alpha|}}{\partial x^\alpha} (D^\beta T_y(\varphi(x,y))) \right)$$

$$= (-1)^{|\alpha|} S_x \left( D^\beta T_y \left( \frac{\partial^{|\alpha|}}{\partial x^\alpha} \varphi(x,y) \right) \right)$$

$$= D^\alpha S_x (D^\beta T_y(\varphi(x,y)))$$

$$= (D^\alpha S \otimes D^\beta T)\varphi(\cdot,\cdot).$$

Thus, the proof is finished. □

THEOREM 3.4.2. *If $S$ and $T$ are in $\mathcal{D}'(\mathbb{R}^m)$ and $\mathcal{D}'(\mathbb{R}^n)$ respectively, then we have*

$$\text{supp}(S \otimes T) = \text{supp}\, S \times \text{supp}\, T. \tag{3.4.2}$$

PROOF. First of all we show that $\text{supp}\, S \times \text{supp}\, T \subset \text{supp}(S \otimes T)$. Indeed, suppose that $(x_0, y_0) \in \text{supp}\, S \times \text{supp}\, T$ and that $U(x_0, y_0)$ is an open neighbourhood of $(x_0, y_0)$ in $\mathbb{R}^m \times \mathbb{R}^n$, then there exist open neighbourhoods $U_1(x_0)$ of $x_0$ in $\mathbb{R}^m$ and $U_2(y_0)$ of $y_0$ in $\mathbb{R}^n$ such that $U_1(x_0) \times U_2(y_0) \subset U(x_0, y_0)$. Take the two functions $\varphi_1 \in \mathcal{D}(U_1(x_0))$ and $\varphi_2 \in \mathcal{D}(U_2(y_0))$ so that $S(\varphi_1) T(\varphi_2) \neq 0$. Of course, $(S \otimes T)(\varphi_1 \otimes \varphi_2) \neq 0$. Therefore $\text{supp}\, S \times \text{supp}\, T \subset \text{supp}(S \otimes T)$.

We shall now show the converse inclusion. Let $\varphi(\cdot, \cdot)$ be in $\mathcal{D}(\mathbb{R}^m \times \mathbb{R}^n)$ and suppose that $\text{supp}\, \varphi \cap (\text{supp}\, S \times \text{supp}\, T) = \emptyset$. Then there exists an open set $U$ in $\mathbb{R}^m$ such that $\text{supp}\, S \subset U$ and $\text{supp}\, \varphi(x, \cdot) \cap \text{supp}\, T = \emptyset$ for $x \in U$. Therefore

$$\psi(x) := T(\varphi(x, \cdot)) = 0$$

for $x \in U$. Thus, we have $S(\psi) = S_x(T_y(\varphi(x,y))) = 0$. It means that $(S \otimes T)\varphi(\cdot, \cdot)$ vanishes for $\varphi$ in $\mathcal{D}(\mathbb{R}^m \times \mathbb{R}^n) - (\text{supp}\, S \times \text{supp}\, T)$. This statement finishes the proof. □

## 3.5. The theorem of kernels

Let $g$ be in $C(\mathbb{R}^m \times \mathbb{R}^n)$. Put by definition

$$(\mathcal{K}\varphi)(x) := \int_{\mathbb{R}^n} g(x,y)\varphi(y)dy$$

for $\varphi \in \mathcal{D}(\mathbb{R}^n)$. It is easy to verify that the mapping $\varphi \to \mathcal{K}\varphi$ from $\mathcal{D}(\mathbb{R}^n)$ into $C(\mathbb{R}^m)$ is continuous. We will consider the bilinear form $\langle \mathcal{K}\varphi, \psi \rangle$ on $\mathcal{D}(\mathbb{R}^n) \times \mathcal{D}(\mathbb{R}^m)$ defined by

$$\langle \mathcal{K}\varphi, \psi \rangle := \int_{\mathbb{R}^m} (\mathcal{K}\varphi)(x)\psi(x)\, dx = \int_{\mathbb{R}^m} \int_{\mathbb{R}^n} g(x,y)\varphi(y)\psi(x)\, dy\, dx, \tag{3.5.1}$$

where $\varphi \in \mathcal{D}(\mathbb{R}^n)$ and $\psi \in \mathcal{D}(\mathbb{R}^m)$. It is clear that this bilinear form is continuous. Simultaneously we can observe that the mapping

$$\varphi \otimes \psi \to \langle \mathcal{K}\varphi, \psi \rangle \tag{3.5.2}$$

is the restriction of the regular distribution $K_g$ generated by $g$ on the set of all functions $\varphi \otimes \psi$, where $\varphi \in \mathcal{D}(\mathbb{R}^n)$ and $\psi \in \mathcal{D}(\mathbb{R}^m)$. These statements give us the following relation

$$\langle \mathcal{K}\varphi, \psi \rangle = K_g(\varphi \otimes \psi) \tag{3.5.3}$$

for $\varphi \in \mathcal{D}(\mathbb{R}^n)$ and $\psi \in \mathcal{D}(\mathbb{R}^m)$. One can ask whether for every linear continuous operator $\mathcal{K} : \mathcal{D}(\mathbb{R}^n) \to \mathcal{D}'(\mathbb{R}^m)$ there exists a distribution $K \in \mathcal{D}'(\mathbb{R}^{m+n})$ such that

$$\langle \mathcal{K}\varphi, \psi \rangle = K(\varphi \otimes \psi) \tag{3.5.4}$$

for $\varphi \in \mathcal{D}(\mathbb{R}^n)$ and $\psi \in \mathcal{D}(\mathbb{R}^m)$, where $\langle \mathcal{K}\varphi, \psi \rangle$ denotes the value of $\mathcal{K}\varphi$ for the argument $\psi$. The next theorem will tell us yes.

DEFINITION 3.5.1. We shall call formula (3.5.4) the Schwartz representation of the operator $\mathcal{K}$. The distribution $K$ will be called the kernel of $\mathcal{K}$.

THEOREM 3.5.1 (Schwartz kernel theorem).
  (i) *Every distribution $K$ in $\mathcal{D}'(\mathbb{R}^{m+n})$ determines a linear operator $\mathcal{K}$ : $\mathcal{D}(\mathbb{R}^n) \to \mathcal{D}'(\mathbb{R}^m)$ given by (3.5.4). This operator is continuous in the following sense: if $\varphi_\nu \to 0$ in $\mathcal{D}(\mathbb{R}^n)$ then $\mathcal{K}\varphi_\nu \to 0$ in $\mathcal{D}'(\mathbb{R}^m)$.*
  (ii) *For every continuous operator $\mathcal{K} : \mathcal{D}(\mathbb{R}^n) \to \mathcal{D}'(\mathbb{R}^m)$ there exists exactly one distribution $K$ in $\mathcal{D}'(\mathbb{R}^m \times \mathbb{R}^n)$ so that (3.5.4) holds.*

PROOF OF PART (i). Let $K$ be in $\mathcal{D}'(\mathbb{R}^{m+n})$. It is easily seen that the mapping $\psi \to K(\varphi \otimes \psi)$ is a continuous linear form on $\mathcal{D}(\mathbb{R}^m)$ for fixed $\varphi$ in $\mathcal{D}(\mathbb{R}^n)$. We are now in a position to define a linear operator $\mathcal{K}$ from $\mathcal{D}(\mathbb{R}^n)$ into $\mathcal{D}'(\mathbb{R}^m)$ taking $\mathcal{K}\varphi(\psi) = K(\varphi \otimes \psi)$. We have to show that the mapping $\varphi \to \mathcal{K}\varphi$ from $\mathcal{D}(\mathbb{R}^n)$ into $\mathcal{D}'(\mathbb{R}^m)$ is continuous. In order to do it we have to verify that if $\varphi_\nu \to 0$ in $\mathcal{D}(\mathbb{R}^n)$, then for fixed $\psi$ in $\mathcal{D}(\mathbb{R}^m)$, $K(\varphi_\nu \otimes \psi) \to 0$, too. This fact is evident. This finishes the proof of part (i). $\qquad \square$

Note that the mapping

$$\mathcal{D}(\mathbb{R}^n) \times \mathcal{D}(\mathbb{R}^m) \ni (\varphi, \psi) \to K(\varphi \otimes \psi)$$

is a bilinear continuous form (see Theorem 1.1.1 and Appendix).

PROOF OF PART (ii). We shall now prove that for every continuous linear operator $\mathcal{K} : \mathcal{D}(\mathbb{R}^n) \to \mathcal{D}'(\mathbb{R}^m)$ there exists exactly one distribution $K$ in $\mathcal{D}'(\mathbb{R}^{m+n})$ such that (3.5.4) holds. Analysis similar to that of the proof of Theorem 3.3.1 shows that there exists at most one distribution $K$ in $\mathcal{D}'(\mathbb{R}^{m+n})$ satisfying (3.5.4). Therefore, it remains to prove the existence of such distribution. It is easy to see that the mapping

$$\mathcal{D}(\mathbb{R}^n) \times \mathcal{D}(\mathbb{R}^m) \ni (\varphi, \psi) \to \langle \mathcal{K}\varphi, \psi \rangle$$

is bilinear and separately continuous. Let $\Omega_1 \Subset \mathbb{R}^m$ and $\Omega_2 \Subset \mathbb{R}^n$. Let $C_0^\infty(\Omega_1)$ and $C_0^\infty(\Omega_2)$ be defined as follows

$$C_0^\infty(\Omega_1) = \{\varphi \in C^\infty(\Omega_1) : \partial^\alpha \varphi \in C_0(\Omega_1), \ \alpha \in \mathbb{N}^m\}$$

and

$$C_0^\infty(\Omega_2) = \{\varphi \in C^\infty(\Omega_2) : \ \partial^\alpha \varphi \in C_0(\Omega_2), \ \alpha \in \mathbb{N}^n\}.$$

Of course, the sets $C_0^\infty(\Omega_1)$ and $C_0^\infty(\Omega_2)$ are vector spaces. Let us equip these vector spaces with the norms $\|\cdot\|_{m,\overline{\Omega}_1}$ and $\|\cdot\|_{m,\overline{\Omega}_2}$, $m = 0, 1, 2, \dots$, respectively (see Section 1.1). $C_0^\infty(\Omega_1)$ and $C_0^\infty(\Omega_2)$ are easily seen to be complete. By virtue of Theorem 2.7.1 we infer that $\mathcal{D}(\Omega_1)$ and $\mathcal{D}(\Omega_2)$ are dense in $C_0^\infty(\Omega_1)$ and $C_0^\infty(\Omega_2)$, respectively. Every linear form $\Lambda \in \mathcal{D}'(\Omega_1)$ may be by continuity prolonged from $\mathcal{D}(\Omega_1)$ onto $C_0^\infty(\Omega_1)$. Thus, we can regard the operator $\mathcal{K} : \mathcal{D}(\mathbb{R}^m) \to \mathcal{D}'(\mathbb{R}^n)$ as a linear continuous operator from $\mathcal{D}(\Omega_2)$ into $(C_0^\infty(\Omega_1))'$, where $(C_0^\infty(\Omega_1))'$ consists of all linear continuous forms on $C_0^\infty(\Omega_1)$. Since $(C_0^\infty(\Omega_1))'$ is sequentially complete (see Th.A.1), therefore the operator $\mathcal{K} : \mathcal{D}(\Omega_2) \to (C_0^\infty(\Omega_1))'$ may be extended by continuity from $\mathcal{D}(\Omega_2)$ onto $C_0^\infty(\Omega_2)$. This extension will be denoted by $\tilde{\mathcal{K}}$. $\langle \tilde{\mathcal{K}} \cdot, \cdot \rangle$ is easily seen to be a bilinear form on $C_0^\infty(\Omega_1) \times C_0^\infty(\Omega_2)$. It is simple matter to show that $\langle \tilde{\mathcal{K}} \cdot, \cdot \rangle$ is separately continuous. This implies that it is jointly continuous, too (see Theorem A.3). Therefore there exist non-negative integers $m_1$, $m_2$ and $C > 0$ such that

$$|\langle \mathcal{K}\varphi, \psi \rangle| \leq C \|\varphi\|_{m_2,\overline{\Omega}_2} \|\psi\|_{m_1,\overline{\Omega}_1}, \tag{3.5.5}$$

if $\psi \in \mathcal{D}(\Omega_1)$ and $\varphi \in \mathcal{D}(\Omega_2)$.

Take two compact sets $X \subset \Omega_1$ and $Y \subset \Omega_2$ and assume that $\operatorname{dist}(X, \partial\Omega_1) > \epsilon_0$ and $\operatorname{dist}(Y, \partial\Omega_2) > \epsilon_0$. Let $\psi_1$ and $\psi_2$ be in $\mathcal{D}(\mathbb{R}^m)$ and $\mathcal{D}(\mathbb{R}^n)$, respectively. Further, we require that $\operatorname{supp}\psi_1 \subset B_1^m(0) \subset \mathbb{R}^m$, $\operatorname{supp}\psi_2 \subset B_1^n(0) \subset \mathbb{R}^n$, $\int_{\mathbb{R}^m}\psi_1 = \int_{\mathbb{R}^n}\psi_2 = 1$. Put by definition

$$K_\epsilon(x, y) := \epsilon^{-m-n} \left\langle \mathcal{K}\psi_2 \left(\frac{y - \cdot}{\epsilon}\right), \psi_1\left(\frac{x - \cdot}{\epsilon}\right) \right\rangle. \tag{3.5.6}$$

It is easily seen that $\psi_2\left(\frac{y-\cdot}{\epsilon}\right) \in \mathcal{D}(\Omega_2)$ and $\psi_1\left(\frac{x-\cdot}{\epsilon}\right) \in \mathcal{D}(\Omega_1)$ if $y \in Y$, $x \in X$ and $\epsilon < \epsilon_0$. In view of (3.5.5) we have

$$|K_\epsilon(x, y)| \leq C \left\|\epsilon^{-n}\psi_2\left(\frac{y - \cdot}{\epsilon}\right)\right\|_{m_2,\overline{\Omega}_2} \left\|\epsilon^{-m}\psi_1\left(\frac{x - \cdot}{\epsilon}\right)\right\|_{m_1,\overline{\Omega}_1} \tag{3.5.7}$$

if $x \in X$ and $y \in Y$. An easy computation shows that $K_{(\cdot)}(\cdot, \cdot)$ is continuous on

$$A = X \times Y \times \{\epsilon : 0 < \epsilon < \epsilon_0\}.$$

Moreover, this continuity is uniform under the variables $x$ and $y$. Suppose that there exists a distribution $K$ in $\mathcal{D}'(\mathbb{R}^{m+n})$ fulfilling (3.5.4), then equality (3.5.6) may be written in the following convolution form

$$K_\epsilon(x, y) = (K * \psi_\epsilon)(x, y), \tag{3.5.8}$$

where

$$\psi_\epsilon(x, y) = \epsilon^{-(m+n)}\psi_1\left(\frac{x}{\epsilon}\right)\psi_2\left(\frac{y}{\epsilon}\right)$$

(see Section 3.3). The above equality indicates to us a way of determining the required distribution $K$. We only need to show that $K_\epsilon(\cdot, \cdot)$ converges to a distribution $K$ in $\mathcal{D}'(\mathbb{R}^{m+n})$ and that it fulfils (3.5.4). A standard computation gives us the following equalities

$$
\frac{\partial}{\partial \epsilon} K_\epsilon(x, y) = \left\langle \mathcal{K}\left(\frac{\partial}{\partial \epsilon}\left(\epsilon^{-n}\psi_2\left(\frac{y-\cdot}{\epsilon}\right)\right)\right), \epsilon^{-m}\psi_1\left(\frac{x-\cdot}{\epsilon}\right) \right\rangle
$$

$$
+ \left\langle \mathcal{K}\left(\epsilon^{-n}\psi_2\left(\frac{y-\cdot}{\epsilon}\right)\right), \frac{\partial}{\partial \epsilon}\left(\epsilon^{-m}\psi_1\left(\frac{x-\cdot}{\epsilon}\right)\right) \right\rangle, \tag{3.5.9}
$$

$$
\frac{\partial}{\partial y_\nu} K_\epsilon(x, y) = \left\langle \mathcal{K}\left(\frac{\partial}{\partial y_\nu}\left(\epsilon^{-n}\psi_2\left(\frac{y-\cdot}{\epsilon}\right)\right)\right), \epsilon^{-m}\psi_1\left(\frac{x-\cdot}{\epsilon}\right) \right\rangle, \tag{3.5.10}
$$

where $y = (y_1, \ldots, y_n)$ and $\nu = 1, \ldots, n$,

$$
\frac{\partial}{\partial x_\mu} K_\epsilon(x, y) = \left\langle \mathcal{K}\left(\epsilon^{-n}\psi_2\left(\frac{y-\cdot}{\epsilon}\right)\right), \frac{\partial}{\partial x_\mu}\left(\epsilon^{-m}\psi_1\left(\frac{x-\cdot}{\epsilon}\right)\right) \right\rangle, \tag{3.5.11}
$$

where $x = (x_1, \ldots, x_m)$ and $\mu = 1, \ldots, m$. Of course, the functions $\partial_\epsilon K_{(\cdot)}(\cdot, \cdot)$, $\partial_{y_\nu} K_{(\cdot)}(\cdot, \cdot)$ and $\partial_{x_\mu} K_{(\cdot)}(\cdot, \cdot)$ are continuous in $A$. One can show that $K_{(\cdot)}(\cdot, \cdot)$ is infinitely differentiable in $A$. Taking into account (3.5.5), one can show that

$$
|K_\epsilon(x, y)| < M\epsilon^{-s} \tag{3.5.12}
$$

if $(x, y) \in X \times Y$, where $s = m + n + m_1 + m_2$ (compare (2.5.3)). We shall now prove the crucial equality for our proof

$$
\frac{\partial}{\partial \epsilon}\left(\epsilon^{-n}\psi\left(\frac{x}{\epsilon}\right)\right) = \sum_{j=1}^{n} \frac{\partial}{\partial x_j}\left(\epsilon^{-n}\psi_j\left(\frac{x}{\epsilon}\right)\right), \tag{3.5.13}
$$

for $\psi \in C^{(1)}(\mathbb{R}^n)$, where $\psi_j(x) = -x_j\psi(x)$. Note that

$$
\partial^{e_j}\psi_j(x) = -\psi(x) - x_j\partial^{e_j}\psi(x).
$$

Hence we get

$$
\partial^{e_j}\psi_j\left(\frac{x}{\epsilon}\right) = -\psi\left(\frac{x}{\epsilon}\right) - \frac{x_j}{\epsilon}\partial^{e_j}\psi\left(\frac{x}{\epsilon}\right). \tag{3.5.14}
$$

On the other hand we have

$$
\frac{\partial}{\partial \epsilon}\left(\epsilon^{-n}\psi\left(\frac{x}{\epsilon}\right)\right) = -n\epsilon^{-n-1}\psi\left(\frac{x}{\epsilon}\right) - \epsilon^{-n}\sum_{\nu=1}^{n}\partial^{e_\nu}\psi\left(\frac{x}{\epsilon}\right)\frac{x_\nu}{\epsilon^2}
$$

$$
= -n\epsilon^{-n-1}\psi\left(\frac{x}{\epsilon}\right) - \epsilon^{-n-1}\sum_{\nu=1}^{n}\partial^{e_\nu}\psi\left(\frac{x}{\epsilon}\right)\frac{x_\nu}{\epsilon}.
$$

From this by (3.5.14) we obtain

$$
\frac{\partial}{\partial \epsilon}\left(\epsilon^{-n}\psi\left(\frac{x}{\epsilon}\right)\right) = -n\epsilon^{-n-1}\psi\left(\frac{x}{\epsilon}\right) + \epsilon^{-n-1}\sum_{\nu=1}^{n}\left(\psi\left(\frac{x}{\epsilon}\right) + \partial^{e_\nu}\psi_\nu\left(\frac{x}{\epsilon}\right)\right)
$$

$$
= \epsilon^{-n-1}\sum_{\nu=1}^{n}\partial^{e_\nu}\psi_\nu\left(\frac{x}{\epsilon}\right) = \sum_{\nu=1}^{n}\frac{\partial}{\partial x_\nu}\left(\epsilon^{-n}\psi_\nu\left(\frac{x}{\epsilon}\right)\right).
$$

Thus, (3.5.13) is proved. In view of (3.5.13), equality (3.5.9) may be written as follows

$$\frac{\partial}{\partial \epsilon} K_\epsilon(x,y) = \left\langle \mathcal{K}\left( \sum_{\nu=1}^n \frac{\partial}{\partial y_\nu}\left( \epsilon^{-n}\psi_{2,(0,\nu)}\left(\frac{y - \cdot}{\epsilon}\right) \right) \right), \epsilon^{-m}\psi_1\left(\frac{x - \cdot}{\epsilon}\right) \right\rangle$$
$$+ \left\langle \mathcal{K}\left( \epsilon^{-n}\psi_2\left(\frac{y - \cdot}{\epsilon}\right) \right), \sum_{\mu=1}^m \frac{\partial}{\partial x_\mu}\left( \epsilon^{-m}\psi_{1,(\mu,0)}\left(\frac{x - \cdot}{\epsilon}\right) \right) \right\rangle,$$

where $\psi_{2,(0,\nu)}(y) = -y_\nu\psi_2(y)$ and $\psi_{1,(\mu,0)}(x) = -x_\mu\psi_1(x)$. Hence, by (3.5.10) and (3.5.11) we get

$$\frac{\partial}{\partial \epsilon} K_\epsilon(x,y) = \sum_{\nu=1}^n \frac{\partial}{\partial y_\nu} \left\langle \mathcal{K}\left( \epsilon^{-n}\psi_{2,(0,\nu)}\left(\frac{y - \cdot}{\epsilon}\right) \right), \epsilon^{-m}\psi_1\left(\frac{x - \cdot}{\epsilon}\right) \right\rangle$$
$$+ \sum_{\mu=1}^m \frac{\partial}{\partial x_\mu} \left\langle \mathcal{K}\left( \epsilon^{-n}\psi_2\left(\frac{y - \cdot}{\epsilon}\right) \right), \epsilon^{-m}\psi_{1,(\mu,0)}\left(\frac{x - \cdot}{\epsilon}\right) \right\rangle.$$

For simplicity of notations put

$$L_{\epsilon,1,(e_\mu,0)}(x,y) := \left\langle \mathcal{K}\left( \epsilon^{-n}\psi_2\left(\frac{y - \cdot}{\epsilon}\right) \right), \epsilon^{-m}\psi_{1,(\mu,0)}\left(\frac{x - \cdot}{\epsilon}\right) \right\rangle,$$
$$L_{\epsilon,1,(0,e_\nu)}(x,y) := \left\langle \mathcal{K}\left( \epsilon^{-n}\psi_{2,(0,\nu)}\left(\frac{y - \cdot}{\epsilon}\right) \right), \epsilon^{-m}\psi_1\left(\frac{x - \cdot}{\epsilon}\right) \right\rangle.$$

One can observe that these functions satisfy (3.5.12). Finally we have

$$\frac{\partial}{\partial \epsilon} K_\epsilon(x,y) = \sum_{\mu=1}^m \frac{\partial}{\partial x_\mu} L_{\epsilon,1,(e_\mu,0)}(x,y) + \sum_{\nu=1}^n \frac{\partial}{\partial y_\nu} L_{\epsilon,1,(0,e_\nu)}(x,y).$$

Repeating the above algorithm again for the functions $L_{\epsilon,1,(e_\mu,0)}$ and $L_{\epsilon,1,(0,e_\nu)}$ one can show that $\frac{\partial^2}{\partial \epsilon^2} K_\epsilon(x,y)$ is a finite sum of $\frac{\partial^{|(\alpha,\beta)|}}{\partial x^\alpha \partial y^\beta} L_{\epsilon,2,(\alpha,\beta)}$, $|(\alpha,\beta)| = 2$, where the functions $L_{\epsilon,2,(\alpha,\beta)}$ satisfy (3.5.12). Generally, $\frac{\partial^k}{\partial \epsilon^k} K_\epsilon(x,y)$, $k = 0,1,2\ldots$ is a finite sum of $\partial^\alpha \partial^\beta L_{\epsilon,k,(\alpha,\beta)}$, $|(\alpha,\beta)| = k$, which are infinitely differentiable in $A$ and fulfil (3.5.12). We shall now expand the function $K_{(\cdot)}(x,y)$ in the Taylor formula around some point $\delta$, $0 < \delta < \epsilon_0$. Thus, we have

$$K_\epsilon(x,y) = \sum_{j=0}^s \frac{\frac{\partial^j}{\partial \epsilon^j} K_\delta(x,y)}{j!}(\epsilon - \delta)^j + \frac{(\epsilon - \delta)^{s+1}}{s!}\int_0^1 \frac{\partial^{s+1}}{\partial \epsilon^{s+1}} K_{\delta + t(\epsilon - \delta)}(x,y)(1-t)^s \, dt.$$

Assume that $\varphi \in \mathcal{D}(\mathbb{R}^{m+n})$, $\operatorname{supp}\varphi \subset X \times Y$, then we have

$$\int_X \int_Y K_\epsilon(x,y)\varphi(x,y) \, dx \, dy = \sum_{j=0}^s \frac{(\epsilon - \delta)^j}{j!} \int_X \int_Y \frac{\partial^j}{\partial \epsilon^j} K_\delta(x,y)\varphi(x,y) \, dx \, dy$$

$$+ \frac{(\epsilon - \delta)^{s+1}}{s!} \int_0^1 \int_X \int_Y \frac{\partial^{s+1}}{\partial \epsilon^{s+1}} K_{\delta + t(\epsilon - \delta)}(x,y)\varphi(x,y)(1-t)^s \, dx \, dy \, dt.$$

Since $\frac{\partial^{s+1}}{\partial \epsilon^{s+1}} K_{(\cdot)}(\cdot, \cdot)$ is a finite sum of $\partial^{\alpha} \partial^{\beta} L_{\epsilon, s+1, (\alpha, \beta)}(\cdot, \cdot)$, where $|(\alpha, \beta)| = s + 1$ we only need to show that there exists

$$\lim_{\epsilon \to 0} \frac{1}{s!} \int\limits_0^1 \int\limits_X \int\limits_Y (\epsilon - \delta)^{s+1} L_{\delta + t(\epsilon - \delta), s+1, (\alpha, \beta)}(x, y)(1 - t)^s \varphi(x, y) \, dx \, dy \, dt =: M(\varphi)$$

for $\varphi \in \mathcal{D}(\mathbb{R}^{m+n})$, where $\operatorname{supp} \varphi \subset X \times Y$. By (3.5.12) we get

$$|(\epsilon - \delta)^{s+1} L_{\delta + t(\epsilon - \delta), s+1, (\alpha, \beta)}(x, y)(1 - t)^s \varphi(x, y)|$$
$$\leq C \delta^{-s} (\delta - \epsilon)^{s+1} \leq C \left(1 - \frac{\epsilon}{\delta}\right)^s (\delta - \epsilon) < C \delta$$

for $O < \epsilon < \delta \leq 1$ and $0 \leq t \leq 1$. Since $L_{(\cdot), s+1, (\alpha, \beta)}(\cdot, \cdot)$ is continuous in $A$, therefore by the Lebesgue dominated convergence theorem we have

$$M(\varphi) = \frac{(-\delta)^{s+1}}{s!} \int\limits_0^1 \int\limits_X \int\limits_Y L_{\delta(1-t), s+1, (\alpha, \beta)}(x, y)(1 - t)^s \varphi(x, y) \, dx \, dy \, dt.$$

Finally we obtain

$$\lim_{\epsilon \to 0} \int\limits_X \int\limits_Y K_{\epsilon}(x, y)\varphi(x, y) \, dx \, dy = \sum_{j=0}^s \frac{(-\delta)^j}{j!} \int\limits_X \int\limits_Y \frac{\partial^j}{\partial \epsilon^j} K_{\delta}(x, y)\varphi(x, y) \, dx \, dy$$

$$+ \frac{(-\delta)^{s+1}}{s!} \int\limits_X \int\limits_Y \int\limits_0^1 \frac{\partial^{s+1}}{\partial \epsilon^{s+1}} K_{\delta(1-t)}(x, y)\varphi(x, y)(1 - t)^s \, dx \, dy \, dt.$$

for $\varphi \in \mathcal{D}(\mathbb{R}^{m+n})$ with $\operatorname{supp} \varphi \subset X \times Y$.

Of course, for arbitrary compact sets $X \subset \mathbb{R}^m$ and $Y \subset \mathbb{R}^n$ there exist $\Omega_1 \Subset \mathbb{R}^m$ and $\Omega_2 \Subset \mathbb{R}^n$ such that $X \subset \Omega_1$, $Y \subset \Omega_2$ and (3.5.5) holds (in general $s$ depends on $X$ and $Y$). Since $\mathcal{D}'(\mathbb{R}^{m+n})$ is sequentially complete (Theorem A.2, Example A.1), therefore there exists $K \in \mathcal{D}'(\mathbb{R}^{m+n})$ such that $K_{\epsilon} \to K$ as $\epsilon \to 0$ in $\mathcal{D}'(\mathbb{R}^{m+n})$. We shall now show that (3.5.4) holds for $\varphi \in \mathcal{D}(\mathbb{R}^n)$ and $\psi \subset \mathcal{D}(\mathbb{R}^m)$. Put

$$\psi_{1, \epsilon}(x) := \epsilon^{-m} \psi_1 \left(\frac{-x}{\epsilon}\right)$$

and

$$\psi_{2, \epsilon}(y) = \epsilon^{-n} \psi_2 \left(\frac{-y}{\epsilon}\right).$$

Hence, by (3.5.6)

$$\int\limits_{\mathbb{R}^m} \int\limits_{\mathbb{R}^n} K_{\epsilon}(x, y)\varphi(y)\psi(x) \, dx \, dy = \int\limits_{\mathbb{R}^m} \int\limits_{\mathbb{R}^n} \langle \mathcal{K}(\psi_{2, \epsilon}(\cdot - y)), \psi_{1, \epsilon}(\cdot - x)\rangle \varphi(y)\psi(x) \, dx \, dy.$$

Replacing the integral by Riemann sums and proceeding to limit (see proof of Theorem 3.1.4) we obtain

$$\int\limits_{\mathbb{R}^m} \int\limits_{\mathbb{R}^n} K_{\epsilon}(x, y)\varphi(y)\psi(x) \, dx \, dy = \langle \mathcal{K}(\psi_{2, \epsilon} * \varphi), (\psi_{1, \epsilon} * \psi)\rangle.$$

Since $\langle \mathcal{K}(\cdot), \cdot \rangle$ is continuous on $\mathcal{D}(\mathbb{R}^m) \times \mathcal{D}(\mathbb{R}^n)$, therefore by Lemma 1.2.1 and Theorem 1.3.1 we get

$$\lim_{\epsilon \to 0} \int\limits_{\mathbb{R}^m} \int\limits_{\mathbb{R}^n} K_\epsilon(x, y) \varphi(y) \psi(x) \, dx \, dy = \langle \mathcal{K}(\lim_{\epsilon \to 0}(\psi_{2,\epsilon} * \varphi)), \lim_{\epsilon \to 0}(\psi_{1,2} * \psi) \rangle.$$

Finally we have

$$K(\varphi \otimes \psi) = \langle \mathcal{K}\varphi, \psi \rangle. \qquad \square$$

EXAMPLE 3.5.1. Now we shall find the kernel $K$ of the operator $\mathcal{K}$, $\mathcal{K}(\varphi) = \Lambda * \varphi$, where

$$\Lambda = \sum_{|\beta| \leq m} a_\beta D^\beta \delta, \ a_\beta \in \mathbb{C}.$$

Note that the mapping

$$\mathcal{D}(\mathbb{R}^n \times \mathbb{R}^n) \ni \eta(\cdot, \cdot) \to K(\eta) = \sum_{|\beta| \leq m} a_\beta \int\limits_{\mathbb{R}^n} D^{(0,\beta)} \eta(x, x) \, dx$$

is linear and continuous. Let $\varphi \in \mathcal{D}(\mathbb{R}^n)$, then

$$\partial^\beta \varphi(y) = (D^\beta \delta * \varphi)(y).$$

Put $\eta(x, y) = \psi(x)\varphi(y)$, then we get

$$K(\varphi \otimes \psi) = \sum_{|\beta| \leq m} a_\beta \int\limits_{\mathbb{R}^n} \psi(x)(D^\beta \delta * \varphi)(y) \, dy.$$

We know that $\mathcal{K}\varphi \in C^\infty(\mathbb{R}^n)$ (see Theorem 3.2.1). Thus we obtain

$$K(\varphi \otimes \psi) = \langle \mathcal{K}\varphi, \psi \rangle.$$

For $m = 0$ we get $\mathcal{K}(\varphi) = \delta * \varphi = \varphi$. Therefore, the kernel $K$ of the identity operator is given by formula

$$K(\eta) = \int\limits_{\mathbb{R}^n} \eta(x, x) \, dx$$

for $\eta \in \mathcal{D}(\mathbb{R}^n \times \mathbb{R}^n)$.

## 3.6. Connection between tensor product and convolution product of distributions

Let $f$ and $g$ be in $L^1(\mathbb{R}^n)$. Suppose that $\varphi \in \mathcal{D}(\mathbb{R}^n)$, then an easy computation shows that

$$\int\limits_{\mathbb{R}^n} \left( \int\limits_{\mathbb{R}^n} f(x - y)g(y)dy \right) \varphi(x) \, dx = \int\limits_{\mathbb{R}^n} \int\limits_{\mathbb{R}^n} f(x)g(y)\varphi(x + y) \, dx \, dy.$$

This equality may be written as follows

$$\int\limits_{\mathbb{R}^n} (f * g)(x)\varphi(x) \, dx = (f_x \otimes g_y)\varphi(x + y). \qquad (3.6.1)$$

Let $S$ and $T$ be in $\mathcal{D}'(\mathbb{R}^n)$. Assume that $\varphi \in \mathcal{D}(\mathbb{R}^n)$. If the expression $(S_x \otimes T_y)\varphi(x+y)$ is sensible for each $\varphi$ in $\mathcal{D}(\mathbb{R}^n)$, then equality (3.6.1) justifies taking the following equality

$$(S * T)(\varphi) := (S_x \otimes T_y)\varphi(x+y) \qquad (3.6.2)$$

as a definition of the convolution product of $S$ and $T$.

REMARK 3.6.1. In view of (3.6.2) we see at once that if there exist the convolution products $Q * T$ and $S * T$, then there exist the convolution products $(Q+S) * T$ and $(Q+S) * T = Q * T + S * T$.

EXAMPLE 3.6.1. Put $S = T = \delta$, and then we have

$$(\delta * \delta)(\varphi) = (\delta_x \otimes \delta_y)\varphi(x+y) = \varphi(0) = \delta(\varphi).$$

We shall now give another example.

EXAMPLE 3.6.2. Let $S$ be an arbitrary distribution and $T$ be a distribution with a compact support. We shall now show that $(S_x \otimes T_y)\varphi(x+y)$ is meaningful for each $\varphi$ in $\mathcal{D}(\mathbb{R}^n)$. In order to do it take by definition $\psi(x) := T_y(\varphi(x+y))$. We have to show that $\psi$ is in $\mathcal{D}(\mathbb{R}^n)$. One can verify that

$$\operatorname{supp} \psi \subset \overline{-\operatorname{supp} T + \operatorname{supp} \varphi} \Subset \mathbb{R}^n.$$

Note that $\partial^\beta \varphi(x + \cdot)$ tends to $\partial^\beta \varphi(x_0 + \cdot)$ as $x \to x_0$ in $C^\infty(\mathbb{R}^n)$. Therefore $\psi$ is continuous in $\mathbb{R}^n$. An easy computation shows that

$$\frac{\partial^{|\alpha|}}{\partial x^\alpha}\psi(x) = T_y\left(\frac{\partial^{|\alpha|}}{\partial x^\alpha}\varphi(x+y)\right).$$

From the above consideration, it follows that $\psi$ is in $\mathcal{D}(\mathbb{R}^n)$. Thus, if $S$ is an arbitrary distribution and $T$ has a compact support, then the convolution product $S * T$ may be defined by (3.6.2). Since the mapping

$$\mathcal{D}(\mathbb{R}^n) \ni \varphi \to \varphi(\cdot + \cdot) \in C^\infty(\mathbb{R}^n \times \mathbb{R}^n)$$

is continuous, therefore $S * T \in \mathcal{D}'(\mathbb{R}^n)$.

THEOREM 3.6.1. If $S \in \mathcal{D}'(\mathbb{R}^n)$ and $\alpha \in \mathcal{D}(\mathbb{R}^n)$, then the convolution product $S * \alpha$ in the above sense agrees with the regularization $S_\alpha$ of $S$ (see Definition 3.1.1).

PROOF. In accordance with Definition 3.3.1 we have

$$(S * \alpha)(\varphi) = (S_x \otimes \alpha_y)\varphi(x+y) = S_x\left(\int\limits_{\mathbb{R}^n} \alpha(y)\varphi(x+y)dy\right)$$

$$= S_x\left(\int\limits_{\mathbb{R}^n} \alpha(w-x)\varphi(w)dw\right) = (S_x \otimes 1_w)(\alpha(w-x)\varphi(w)),$$

where $1(w) = 1$, $w \in \mathbb{R}^n$. By commutativity of tensor product we infer that

$$(S_x \otimes \alpha_j)\varphi(x+y) = (1_w \otimes S_x)(\alpha(w-x)\varphi(w)) = \int\limits_{\mathbb{R}^n} \varphi(w)S_x(\alpha(w-x))\, dw.$$

Therefore $(S * \alpha)(\varphi) = S_\alpha(\varphi)$. This statement finishes the proof. $\square$

In general, equality (3.6.2) is not sensible, because $\varphi(\cdot + \cdot)$ does not belong to $\mathcal{D}(\mathbb{R}^n \times \mathbb{R}^n)$. For sensibility of (3.6.2) we have to take some restrictions concerning the supports of $S$ and $T$.

DEFINITION 3.6.1. If for each bounded set $K \subset \mathbb{R}^n$ the set $A \cap (K - B)$ is also bounded, then we say that the sets $A$ and $B$ are regular at infinity.

EXAMPLE 3.6.3.
  (i) If $A$ is an arbitrary set in $\mathbb{R}^n$ and $B$ is bounded, then $A$ and $B$ are regular at infinity.
  (ii) The sets $A = \{x : x \in \mathbb{R},\ x \geq a\}$ and $B = \{x : x \in \mathbb{R},\ x \geq b\}$ are regular at infinity.
  (iii) The sets $A = \{(x, y) : (x, y) \in \mathbb{R}^2,\ x \geq 0\}$ and $B = \{(x, y) : (x, y) \in \mathbb{R}^2,\ x \geq 0,\ |y| < x\}$ are regular at infinity.
  (iv) The sets $A = B = \{(x, y) : (x, y) \in \mathbb{R}^2,\ x \geq 0\}$ are not regular at infinity.

We begin with presenting a lemma.

LEMMA 3.6.1. *If any sets $A$ and $B$ are regular at infinity and $K$ is bounded, then the set*
$$\Omega_K = \{(x, y) : x \in A,\ y \in B \text{ and } x + y \in K\}$$
*is bounded in $\mathbb{R}^n \times \mathbb{R}^n$.*

PROOF. Since the sets $A$ and $B$ are regular at infinity, therefore the set $C = A \cap (K - B)$ is bounded for every bounded $K \subset \mathbb{R}^n$. Fix $K$, let $x \in A$, $z \in K$ and $y \in B$, then $x = z - y \in C$. Note that for all pairs $(x, y)$ in $\Omega_K$, their first coordinates are bounded. Since $K$ is bounded, the second coordinates of $(x, y) \in \Omega_K$ are also bounded. This finishes the proof.          $\square$

Let $A_\varphi = \{(x, y) : x + y \in \operatorname{supp} \varphi\}$. Put $\Omega_\varphi := A_\varphi \cap (\operatorname{supp} S \times \operatorname{supp} T)$. It is easily seen that the sets $A_\varphi$ and $\operatorname{supp} S \times \operatorname{supp} T$ are closed in $\mathbb{R}^{2n}$. Taking into account Theorem 3.4.2 and Lemma 3.6.1 we infer that the following two cases

$$\Omega_\varphi \text{ is a non-empty compact set,} \tag{3.6.3}$$

$$\operatorname{dist}(A_\varphi, \operatorname{supp}(S \otimes T)) > 0 \tag{3.6.4}$$

hold. We are now in a position to give a definition of the convolution product of two distributions having regular supports at infinity. Suppose now that (3.6.3) holds. Therefore, by Lemma 2.1.1 there exists a function $\psi_\varphi$ in $\mathcal{D}(\mathbb{R}^{2n})$ such that $\psi_\varphi(x, y) = 1$ for $(x, y) \in \overline{B_\epsilon(\Omega_\varphi)}$, $\epsilon > 0$. Of course, $\varphi(\cdot + \cdot)\psi_\varphi(\cdot, \cdot)$ is in $\mathcal{D}(\mathbb{R}^{2n})$. If (3.6.4) is satisfied, then for any $\Theta \in \mathcal{D}(\mathbb{R}^{2n})$ such that $\operatorname{supp} \Theta \cap \operatorname{supp}(S \otimes T) = \emptyset$ we have $(S \otimes T)(\Theta \varphi(\cdot + \cdot)) = 0$ In this case, it is natural to take $\psi_\varphi = 0$. Put

$$(S * T)(\varphi) := (S_x \otimes T_y)(\varphi(x + y)\psi_\varphi(x, y)) \tag{3.6.5}$$

for $\varphi \in \mathcal{D}(\mathbb{R}^n)$. Of course, $(S * T)(\varphi) = 0$ when (3.6.4) holds. (Compare with Section 2.4.)

THEOREM 3.6.2. *The mapping*

$$\varphi \to (S * T)(\varphi) \tag{3.6.6}$$

*given by (3.6.5) is a linear continuous form on $\mathcal{D}(\mathbb{R}^n)$.*

PROOF. We have to show that the right side of (3.6.5) does not depend on a choice of the function $\psi_\varphi$. Let $\psi_\varphi^1$ and $\psi_\varphi^2$ be in $\mathcal{D}(\mathbb{R}^{2n})$ and

$$\psi_\varphi^1(x, y) = \psi_\varphi^2(x, y) = 1$$

for $(x, y) \in \overline{B_\epsilon(\Omega_\varphi)}$. Hence we infer that

$$\operatorname{supp}(S \otimes T) \cap \operatorname{supp}((\psi_\varphi^1 - \psi_\varphi^2)\varphi(\cdot + \cdot)) = \emptyset.$$

Therefore

$$(S \otimes T)((\psi_\varphi^1 - \psi_\varphi^2)\varphi(\cdot + \cdot)) = 0.$$

This means that the right side of (3.6.5) does not depend on $\psi_\varphi$. We shall now show the linearity of (3.6.5). Put

$$\Omega_{\varphi_1, \varphi_2} := \{(x, y) : x \in \operatorname{supp} S, \ y \in \operatorname{supp} T, \ x + y \in \operatorname{supp} \varphi_1 \cup \operatorname{supp} \varphi_2\}.$$

Moreover, assume that $\psi_{\varphi_1, \varphi_2} \in \mathcal{D}(\mathbb{R}^{2n})$ and $\psi_{\varphi_1, \varphi_2}(x, y) = 1$ for $(x, y) \in \overline{B_\epsilon(\Omega_{\varphi_1, \varphi_2})}$. Note that

$$(S_x \otimes T_y)(\varphi_i(x + y)\psi_{\varphi_1, \varphi_2}(x, y)) = (S_x \otimes T_y)(\varphi(x + y)\psi_{\varphi_i}(x, y)), \quad i = 1, 2.$$

The details are left to the reader.

The only thing left to show is that this mapping is continuous. If a sequence $(\varphi_\nu)$ converges to zero in $\mathcal{D}(\mathbb{R}^n)$, then there exists a compact set $K \subset \mathbb{R}^n$ such that $\operatorname{supp} \varphi_\nu \subset K$ for $k = 1, 2, \ldots$ and $(\partial^\alpha \varphi_\nu)$ converges uniformly to zero in $K$ for each $\alpha \in \mathbb{N}^n$. In view of Lemma 2.1.1 there exists a function $\psi_{\Omega_K}$ in $\mathcal{D}(\mathbb{R}^{2n})$ such that $\psi_{\Omega_K}(x, y) = 1$ for $(x, y) \in \overline{B_\epsilon(\Omega_K)}$. It is evident that the sequence $(\psi_{\Omega_K}\varphi_\nu(\cdot + \cdot))$ converges to zero in $\mathcal{D}(\mathbb{R}^{2n})$. Since $S \otimes T$ belongs to $\mathcal{D}'(\mathbb{R}^{2n})$, therefore $S * T$ is in $\mathcal{D}'(\mathbb{R}^n)$. Thus, the proof is finished. $\square$

DEFINITION 3.6.2. The mapping (3.6.6) is called the convolution product of $S$ and $T$.

REMARK 3.6.2. Let the functions $f$ and $g$ be continuous in $\mathbb{R}^n$. Suppose that the supports of $f$ and $g$ are regular at infinity, then integral $\psi(x) = \int_{\mathbb{R}^n} f(x - y)g(y)dy$ is finite for every $x \in \mathbb{R}^n$. Moreover, the function $\psi$ is continuous. It implies that for each $\varphi$ in $\mathcal{D}(\mathbb{R}^n)$ there exist the integrals

$$\int_{\mathbb{R}^n} \left( \int_{\mathbb{R}^n} f(x - y)g(y)dy \right) \varphi(x)dx = \int_{\mathbb{R}^n} g(y) \left( \int_{\mathbb{R}^n} f(x)\varphi(x + y)dx \right) dy$$

$$\int_{\mathbb{R}^n} \int_{\mathbb{R}^n} f(x)g(y)\psi_\varphi(x, y)\varphi(x + y)dxdy,$$

where $\psi_\varphi$ is defined as above. Thus, we have shown that if $f, g \in C(\mathbb{R}^n)$ and their supports are regular at infinity, then there exists the convolution product $f * g$ in the distributional sense.

## 3.7. Differentiation and support of convolution product

In this section we prove two important theorems for applications.

THEOREM 3.7.1. *Let $S$ and $T$ be in $\mathcal{D}'(\mathbb{R}^n)$. If $\operatorname{supp} S$ and $\operatorname{supp} T$ are regular at infinity, then there exists the convolution product $S * T$ and*

$$D^\alpha(S * T) = D^\alpha S * T = S * D^\alpha T \tag{3.7.1}$$

*for each $\alpha \in \mathbb{N}^n$.*

PROOF. The existence of the convolution product was shown in Section 3.6. It remains to prove that (3.7.1) holds. Suppose now that (3.6.3) is satisfied. In accordance with the definition of $S * T$ we have

$$D^{e_i}(S * T)(\varphi) = -(S_x \otimes T_y)(\psi_\varphi(x, y)\partial^{e_i}\varphi(x + y))$$
$$= -S_x\left(T_y\left(\frac{\partial}{\partial y_i}(\psi_\varphi(x, y)\varphi(x + y)) - \frac{\partial}{\partial y_i}\psi_\varphi(x, y)\varphi(x + y)\right)\right).$$

Analysis similar to that in the previous section shows that

$$(S_x \otimes T_y)\left(\frac{\partial}{\partial y_i}\psi_\varphi(x, y)\varphi(x + y)\right) = 0.$$

Therefore

$$D^{e_i}(S * T)(\varphi) = -S_x\left(T_y\left(\frac{\partial}{\partial y_i}(\psi_\varphi(x, y)\varphi(x + y))\right)\right)$$
$$= S_x \otimes (D^{e_i}T)_y(\psi_\varphi(x, y)\varphi(x + y))$$
$$= (S * D^{e_i}T)(\varphi).$$

Equality (3.7.1) is evident when (3.6.4) holds. Similarly, one can show that $D^{e_i}(S * T) = D^{e_i}S * T$. Finally, by induction we have (3.7.1). It proves the theorem. □

THEOREM 3.7.2. *Let $S$ and $T$ be in $\mathcal{D}'(\mathbb{R}^n)$. If $\operatorname{supp} S$ and $\operatorname{supp} T$ are regular at infinity, then*

$$\operatorname{supp}(S * T) \subset \overline{\operatorname{supp} S + \operatorname{supp} T}. \tag{3.7.2}$$

PROOF. It is easy to verify that if $\operatorname{supp}\varphi \cap \overline{\operatorname{supp} S + \operatorname{supp} T} = \emptyset$, then (3.6.4) holds. Hence, $(S * T)(\varphi) = 0$. Thus, the proof is complete. □

CHAPTER 4

# DIFFERENTIAL EQUATIONS

### 4.1. Fundamental solutions of differential equations

To begin with, we shall deal with the problem of the fundamental solution of differential equations with constant coefficients.

DEFINITION 4.1.1. A distribution $E$ which satisfies the differential equation

$$\sum_{|\alpha|\leq m} a_\alpha D^\alpha E = \delta, \qquad (4.1.1)$$

where $a_\alpha \in \mathbb{C}$ is called a fundamental solution of the differential operator $\sum_{|\alpha|\leq m} a_\alpha D^\alpha$.

We are interested in finding a solution of the differential equation

$$\sum_{|\alpha|\leq m} a_\alpha D^\alpha u = f, \qquad (4.1.2)$$

where $f \in \mathcal{D}'(\mathbb{R}^n)$. Suppose that there exists the convolution product $u := E * f$. Then by Theorem 3.7.1 and Remark 3.6.1 we have

$$\sum_{|\alpha|\leq m} a_\alpha D^\alpha u = \left(\sum_{|\alpha|\leq m} a_\alpha D^\alpha E\right) * f = \delta * f = f.$$

Thus, the distribution $u = E * f$ is a solution of Equation (4.1.2).

EXAMPLE 4.1.1. We are looking for a fundamental solution of the differential operator $\square = D^{(0,2)} - D^{(2,0)}$. To this end define a set $\Omega = \{(x,t) : t \geq 0, |x| < t\}$. Set $E = \frac{1}{2}\chi_\Omega$, where $\chi_\Omega$ is the characteristic function of $\Omega$. We prove that $E$ is a fundamental solution of $\square$. We have to show that

$$\iint_{\mathbb{R}^2} \frac{1}{2}\chi_\Omega(x,t) \left(\frac{\partial^2}{\partial t^2}\varphi(x,t) - \frac{\partial^2}{\partial x^2}\varphi(x,t)\right) dx\,dt = \varphi(0,0)$$

for $\varphi \in \mathcal{D}(\mathbb{R}^2)$. This equation can be written in the equivalent form

$$\frac{1}{2}\iint_\Omega \left(\frac{\partial^2}{\partial t^2}\varphi(x,t) - \frac{\partial^2}{\partial x^2}\varphi(x,t)\right) dx\,dt = \varphi(0,0).$$

We shall now change variables $x$ and $t$ in the above integral putting

$$x = \frac{-\xi+\eta}{2} \quad \text{and} \quad t = \frac{\xi+\eta}{2}.$$

An easy computation shows that

$$2\frac{\partial^2}{\partial\xi\partial\eta}\varphi\left(\frac{-\xi+\eta}{2}, \frac{\xi+\eta}{2}\right) = \frac{1}{2}\left(\frac{\partial^2}{\partial t^2}\varphi(x,t) - \frac{\partial^2}{\partial x^2}\varphi(x,t)\right).$$

Therefore

$$\frac{1}{2} \iint\limits_{\Omega} \left( \frac{\partial^2}{\partial t^2} \varphi(x,t) - \frac{\partial^2}{\partial x^2} \varphi(x,t) \right) dx\, dt = \iint\limits_{\Omega'} \frac{\partial^2}{\partial \xi \partial \eta} \varphi \left( \frac{-\xi + \eta}{2}, \frac{\xi + \eta}{2} \right) d\xi\, d\eta,$$

where $\Omega' = \{(\xi, \eta) : \xi \geq 0,\ \eta \geq 0\}$. It is easily seen that

$$\iint\limits_{\Omega'} \frac{\partial^2}{\partial \xi \partial \eta} \varphi \left( \frac{-\xi + \eta}{2}, \frac{\xi + \eta}{2} \right) d\xi\, d\eta = \varphi(0,0),$$

which was supposed to be shown.

We are interested in finding a solution of the differential equation

$$D^{(0,2)} u - D^{(2,0)} u = h, \tag{4.1.3}$$

where $h$ is in $L^1_{loc}(\mathbb{R}^2)$ and $\operatorname{supp} h \subset \mathbb{R} \times \mathbb{R}^+$. It is easily seen that the $\operatorname{supp} E$, $E = \frac{1}{2}\chi_\Omega$, and $\operatorname{supp} h$ are regular at infinity. Therefore by Remark 3.6.2 we infer that the convolution product $E * h$ exists in the classical sense. In fact, we have

$$(E * h)(x,t) = \frac{1}{2} \iint\limits_{\Omega_t} h(x - \xi, t - \eta)\, d\xi\, d\eta,$$

where $\Omega_t = \{(\xi, \eta) : |\xi| < \eta,\ 0 \leq \eta \leq t\}$. Put $\xi' = x - \xi$, $\eta' = t - \eta$, then

$$(E * h)(x,t) = \frac{1}{2} \iint\limits_{\Omega'_t} h(\xi', \eta')\, d\xi'\, d\eta',$$

where $\Omega'_t = -\Omega_t + (x,t)$. Finally we get

$$(E * h)(x,t) = \frac{1}{2} \int_0^t \left( \int_{x-(t-\eta)}^{x+(t-\eta)} h(\xi, \eta)\, d\xi \right) d\eta. \tag{4.1.4}$$

## 4.2. The Cauchy problem for the wave equation with distribution data

The Cauchy problem for the wave equation in one dimensional space seeks for a function $u$ that satisfies

$$\frac{\partial^2}{\partial t^2} u(x,t) - \frac{\partial^2}{\partial x^2} u(x,t) = h(x,t) \tag{4.2.1}$$

for $(x,t) \in \mathbb{R} \times [t_0, \infty)$ and

$$u(x, t_0) = f(x) \quad \text{and} \quad \frac{\partial}{\partial t} u(x, t_0) = g(x) \tag{4.2.2}$$

for $x \in \mathbb{R}$, where $f$ and $g$ are given functions. Let $f$ and $g$ be in $C(\mathbb{R})$. Suppose that a function $u$ belonging to $C^{(2)}(\mathbb{R} \times [t_0, \infty))$ satisfies (4.2.1) and (4.2.2) in the classical sense. A distribution theoretical formulation of the Cauchy problem can be found by beginning with a classical solution $u$ and calculating $D^{(0,2)} u - D^{(2,0)} u$ in the distributional sense. To this end it is natural to extend the definition of $u$ and $h$ from $\mathbb{R} \times [t_0, \infty)$ on all of $\mathbb{R}^2$ by defining $u(x,t) = h(x,t) = 0$ for $(x,y) \in \mathbb{R} \times (-\infty, t_0)$.

With this understanding, one may regard $u$ and $h$ as elements of $\mathcal{D}'(\mathbb{R}^2)$. From the assumption we have

$$\int\limits_{-\infty}^{\infty} \int\limits_{t_0}^{\infty} \left( \frac{\partial^2}{\partial t^2} u(x,t) - \frac{\partial^2}{\partial x^2} u(x,t) \right) \varphi(x,t)\, dx\, dt = \int\limits_{-\infty}^{\infty} \int\limits_{t_0}^{\infty} h(x,t)\varphi(x,t)\, dx\, dt \quad (4.2.3)$$

for $\varphi \in \mathcal{D}(\mathbb{R}^2)$. Taking into account (4.2.2) compute the integral

$$\int\limits_{-\infty}^{\infty} \int\limits_{t_0}^{\infty} \frac{\partial^2}{\partial t^2} u(x,t)\varphi(x,t)\, dx\, dt.$$

Of course,

$$\partial^{(0,1)}(\partial^{(0,1)} u\varphi) = \partial^{(0,2)} u\varphi + \partial^{(0,1)} u \partial^{(0,1)} \varphi.$$

From this we obtain

$$\int\limits_{-\infty}^{\infty} \int\limits_{t_0}^{\infty} \frac{\partial^2}{\partial t^2} u(x,t)\varphi(x,t)\, dx\, dt$$

$$= -\int\limits_{-\infty}^{\infty} \partial^{(0,1)} u(x,t_0)\varphi(x,t_0)\, dx - \int\limits_{-\infty}^{\infty} \int\limits_{t_0}^{\infty} \frac{\partial}{\partial t} u(x,t)\frac{\partial}{\partial t}\varphi(x,t)\, dx\, dt$$

$$= -\int\limits_{-\infty}^{\infty} g(x)\varphi(x,t_0)\, dx - \int\limits_{-\infty}^{\infty} \int\limits_{t_0}^{\infty} \frac{\partial}{\partial t} u(x,t)\frac{\partial}{\partial t}\varphi(x,t)\, dx\, dt.$$

Again note that

$$\partial^{(0,1)}(u\partial^{(0,1)}\varphi) = \partial^{(0,1)} u \partial^{(0,1)}\varphi + u\partial^{(0,2)}\varphi.$$

Hence we get

$$\int\limits_{-\infty}^{\infty} \int\limits_{t_0}^{\infty} \frac{\partial}{\partial t} u(x,t)\frac{\partial}{\partial t}\varphi(x,t)\, dx\, dt$$

$$= -\int\limits_{-\infty}^{\infty} f(x)\frac{\partial}{\partial t}\varphi(x,t_0)\, dx - \int\limits_{-\infty}^{\infty} \int\limits_{t_0}^{\infty} u(x,t)\frac{\partial^2}{\partial t^2}\varphi(x,t)\, dx\, dt.$$

Finally we have

$$\int\limits_{-\infty}^{\infty} \int\limits_{t_0}^{\infty} \frac{\partial^2}{\partial t^2} u(x,t)\varphi(x,t)\, dx\, dt$$

$$= -\int\limits_{-\infty}^{\infty} g(x)\varphi(x,t_0)\, dx + \int\limits_{-\infty}^{\infty} f(x)\frac{\partial}{\partial t}\varphi(x,t_0)\, dx + \int\limits_{-\infty}^{\infty} \int\limits_{t_0}^{\infty} u(x,t)\frac{\partial^2}{\partial t^2}\varphi(x,t)\, dx\, dt.$$

This equality can be written in the following distributional form

$$\int\limits_{-\infty}^{\infty} \int\limits_{t_0}^{\infty} \frac{\partial^2}{\partial t^2} u(x,t) \varphi(x,t) \, dx \, dt$$

$$= -(g_x \otimes \delta_{t_0})\varphi(x,t) - (f_x \otimes D\delta_{t_0})\varphi(x,t) + D^{(0,2)} u(\varphi). \quad (4.2.4)$$

It is easily seen that

$$\int\limits_{-\infty}^{\infty} \int\limits_{t_0}^{\infty} \frac{\partial^2}{\partial x^2} u(x,t) \varphi(x,t) \, dx \, dt = D^{(2,0)} u(\varphi).$$

From this, (4.2.4) and (4.2.3) we obtain

$$D^{(0,2)}u - D^{(2,0)}u = g \otimes \delta_{t_0} + f \otimes D\delta_{t_0} + h =: k. \quad (4.2.5)$$

Thus, we have shown that if $u$ satisfies (4.2.1) and (4.2.2) in the classical sense, then $u$ regarded as a distribution fulfils (4.2.5). Note that $\operatorname{supp} k \subset \mathbb{R} \times [t_0, \infty)$.

Conversely, suppose that $f$, $g$ and $h$ are in $\mathcal{D}'(\mathbb{R})$ and $\operatorname{supp} h \subset \mathbb{R} \times [t_0, \infty)$, then a distribution $u$ satisfying (4.2.5) is called a solution of the Cauchy problem (4.2.1) and (4.2.2) for the distributional data $f$ and $g$. We know that $E = \frac{1}{2}\chi_\Omega$ is a fundamental solution of the wave equation (see Section 4.1). Put $t_0 = 0$ and suppose that $\operatorname{supp} h \subset \mathbb{R}^+ \times [0,\infty)$, then the $\operatorname{supp} E$ and $\operatorname{supp}(f \otimes D\delta + g \otimes \delta + h)$ are regular at infinity. Therefore there exists the convolution product

$$u := E * (f \otimes D\delta + g \otimes \delta + h). \quad (4.2.6)$$

Of course, the distribution $u$ fulfils (4.2.5).

Let $f$ and $g$ be in $C(\mathbb{R})$. We shall now show that (4.2.6) gives us the d'Alembert formula

$$u(x,t) = \frac{1}{2}\left( f(x-t) + f(x+t) + \int\limits_{x-t}^{x+t} g(\eta) \, d\eta + \int\limits_{0}^{t} \int\limits_{x-(t-\eta)}^{x+(t-\eta)} h(\xi,\eta) \, d\xi \, d\eta \right). \quad (4.2.7)$$

Indeed, let $H$ denote the Heaviside step function. Note that

$$E * (f \otimes D\delta) = D^{(0,2)}(E * (f \otimes H)), \quad (4.2.8)$$

$$E * (g \otimes \delta) = D^{(0,1)}(E * (g \otimes H)). \quad (4.2.9)$$

Therefore, we are left with the task of determining the convolution products $E * (f \otimes H)$ and $E * (g \otimes H)$. Let $F$ be a function in $C^{(2)}(\mathbb{R})$ such that $F'' = f$. It is easily seen that

$$(f \otimes H)(x,t) = \begin{cases} f(x) & \text{if } t \geq 0 \\ 0 & \text{if } t < 0. \end{cases}$$

By (4.1.4) we have

$$(E * (f \otimes H))(x,t) = \frac{1}{2}\int\limits_{0}^{t}\left( \int\limits_{x-(t-\eta)}^{x+(t-\eta)} f(\xi) \, d\xi \right) d\eta = \frac{1}{2}(F(x+t) + F(x-t) - 2F(x)).$$

Hence we have

$$\frac{\partial^2}{\partial t^2}(E * (f \otimes H))(x,t) = \frac{1}{2}(f(x+t) + f(x-t)). \quad (4.2.10)$$

Taking $G$ in $C^{(2)}(\mathbb{R})$ so that $G'' = g$, in the same way one can show that

$$(E * (g \otimes H))(x,t) = \frac{1}{2}(G(x+t) + G(x-t) - 2G(x)).$$

Hence

$$\frac{\partial}{\partial t}(E * (g \otimes H))(x,t) = \frac{1}{2}(G'(x+t) - G'(x-t)) = \frac{1}{2}\int\limits_{x-t}^{x+t} g(\eta)\,d\eta. \qquad (4.2.11)$$

Finally, by (4.1.4), (4.2.10) and (4.2.11) we get (4.2.7).

## 4.3. Fundamental solutions of the Laplace operator and the heat operator

We shall consider separately three cases

    (i) $\Delta u = D^2 u$ if $n = 1$,
    (ii) $\Delta u = D^{(2,0)}u + D^{(0,2)}u$ if $n = 2$,
    (iii) $\Delta u = D^{(2,0,\cdots)}u + D^{(0,2,\cdots)}u + \cdots + D^{(0,\cdots,2)}u$ if $n > 2$.

It is easy to verify that $D^2 E = \delta$ if $E(x) = \frac{1}{2}|x|$. We shall now show that the function $E(x) = \frac{1}{2\pi}\ln|x|$ is a fundamental solution for the operator $\Delta = D^{(2,0)} + D^{(0,2)}$. Note that

$$\frac{1}{2\pi}\iint\limits_{B_R(0)} \ln\sqrt{x_1^2 + x_2^2}\,dx_1\,dx_2 = \int\limits_0^R r\ln r\,dr < \infty.$$

Therefore $E$ is in $L^1_{loc}(\mathbb{R}^2)$. We have to show that $\Delta E = \delta$. In order to do this set by definition

$$E_\epsilon(x) := \frac{1}{2\pi}\ln\sqrt{x_1^2 + x_2^2 + \epsilon^2}.$$

Of course, $E_\epsilon$ pointwise converges to $E$ in $\mathbb{R}^2 - \{0\}$ as $\epsilon \to 0$. Moreover

$$0 \le E_\epsilon(x) - E(x) = \frac{1}{2\pi}\ln\sqrt{1 + \frac{\epsilon^2}{|x|^2}}.$$

Suppose that $\epsilon < 1$. Hence we have

$$(E_\epsilon - E)(x) \le (E_1 - E)(x).$$

It is easy to check that $E_1 - E \in L^1_{loc}(\mathbb{R}^2)$. By the Lebesgue dominated convergence theorem we infer that $E_\epsilon$ tends to $E$ as $\epsilon \to 0$ in $L^1_{loc}(\mathbb{R}^2)$. Therefore $E_\epsilon$ converges to $E$ in $\mathcal{D}'(\mathbb{R}^2)$. Since distributional derivation is continuous, what is left is to show that $\Delta E_\epsilon$ tends to $\delta$ in $\mathcal{D}'(\mathbb{R}^2)$. An easy computation shows that

$$\Delta E_\epsilon(x) = \frac{1}{\pi}\epsilon^2(|x|^2 + \epsilon^2)^{-2} = \frac{1}{\pi}\epsilon^{-2}\left(\left|\frac{x}{\epsilon}\right|^2 + 1\right)^{-2}.$$

Put $\psi(x) := \pi^{-1}(|x|^2 + 1)^{-2}$. A standard computation gives $\int_{\mathbb{R}^2}\psi = 1$. In view of Corollary 1.2.1 we infer that $\Delta E_\epsilon$ converges to $\delta$ in $\mathcal{D}'(\mathbb{R}^2)$.

    Using the same method we shall show that the Newtonian potential

$$E(x) = \frac{|x|^{2-n}}{(2-n)\omega_n}, \qquad n > 2,$$

where $\omega_n$ is the area of the unite sphere ($\{x : |x| = 1\}$) in $\mathbb{R}^n$, is a fundamental solution of the Laplace operator $\Delta$. First of all we show that $E$ is in $L^1_{loc}(\mathbb{R}^n)$. To this end compute the integral $\int_{B_R(0)} E(x)dx$. We change variables $x$ by introducing polar variables

$$
\begin{aligned}
x_1 &= r\cos t_1 \cos t_2 \cdots \cos t_{n-2} \cos t_{n-1} \\
x_2 &= r\sin t_1 \cos t_2 \cdots \cos t_{n-2} \cos t_{n-1} \\
x_3 &= r\sin t_2 \cdots \cos t_{n-2} \cos t_{n-1} \\
&\ \ \vdots \qquad\qquad\qquad\qquad \vdots \\
x_n &= r\sin t_{n-1},
\end{aligned}
\tag{4.3.1}
$$

where $0 \le t_1 \le 2\pi$, $-\frac{\pi}{2} \le t_i \le \frac{\pi}{2}$ if $i = 2, \ldots, n-1$. We know that the Jacobian of (4.3.1) equals

$$
J(r, t_1, \ldots, t_{n-1}) = r^{n-1} \cos t_2 \cos^2 t_3 \cdots \cos^{n-2} t_{n-1}
$$

and

$$
\omega_n = \int_0^{2\pi} \int_{-\frac{\pi}{2}}^{\frac{\pi}{2}} \cdots \int_{-\frac{\pi}{2}}^{\frac{\pi}{2}} \cos t_2 \cos^2 t_3 \cdots \cos^{n-2} t_{n-1}\, dt_{n-1} \cdots dt_1.
$$

After the above preliminaries we shall compute the integral $\int_{B_R(0)} E(x)dx$. In fact, we have

$$
\int_{B_R(0)} \frac{1}{(2-n)\omega_n} |x|^{2-n}\, dx = \frac{1}{(2-n)\omega_n} \int_0^R r\omega_n\, dr = \frac{R^2}{2(2-n)}.
$$

Therefore $E$ is in $L^1_{loc}(\mathbb{R}^n)$, $n > 2$. Similarly as in the previous case put

$$
E_\epsilon(x) := \frac{(|x|^2 + \epsilon^2)^{\frac{2-n}{2}}}{(2-n)\omega_n}.
$$

Note that $E$ is in $C^\infty(\mathbb{R}^n)$. Moreover $E_\epsilon(x)$ tends to $E(x)$ as $\epsilon \to 0$ and $-E_\epsilon(x) < -E(x)$ if $x \ne 0$. By the Lebesgue dominated convergence theorem we infer that $E_\epsilon$ converges to $E$ in $L^1_{loc}(\mathbb{R}^n)$. The proof is completed by showing that $\Delta E_\epsilon$ tends to $\delta$ as $\epsilon \to 0$. One can show that

$$
\Delta E_\epsilon(x) = \frac{n\epsilon^{-n}\left(\left|\frac{x}{\epsilon}\right|^2 + 1\right)^{-\frac{n+2}{2}}}{\omega_n}.
\tag{4.3.2}
$$

Set by definition

$$
\psi(x) := \frac{n(|x|^2 + 1)^{-\frac{n+2}{2}}}{\omega_n}.
$$

We shall now show that $\int_{\mathbb{R}^n} \psi(x)dx = 1$. By introducing polar variables in the above integral we obtain

$$
\int_{\mathbb{R}^n} \frac{n(|x|^2 + 1)^{-\frac{n+2}{2}}}{\omega_n}\, dx = \int_0^\infty n(r^2 + 1)^{-\frac{n+2}{2}} r^{n-1}\, dr.
$$

Set $s = r^2 + 1$, then

$$
\int_{\mathbb{R}^n} \psi(x)\, dx = \frac{n}{2} \int_1^\infty \left(1 - \frac{1}{s}\right)^{\frac{n-2}{2}} s^{-2}\, ds.
$$

Putting again $t = \frac{1}{s}$, one can verify that $\int_{\mathbb{R}^n} \psi(x)\, dx = 1$. Taking into account (4.3.2), by Corollary 1.2.1 we conclude that $\Delta E_\epsilon$ converges to $\delta$ in $\mathcal{D}'(\mathbb{R}^n)$.

The same method can be used to find a fundamental solution of the heat operator $D^{(0,\cdots,0,1)} - \Delta$, where $\Delta = D^{(2,,0,\cdots,0,0)} + \cdots + D^{(0,\cdots,2,0)}$. An easy computation shows that

$$(4\pi t)^{-\frac{n}{2}} \int_{\mathbb{R}^n} \exp\left(-\frac{|x|^2}{4t}\right) dx = 1.$$

Put by definition

$$E(x,t) := \begin{cases} (4\pi t)^{-\frac{n}{2}} \exp\left(-\frac{|x|^2}{4t}\right) & \text{if } (x,t) \in \mathbb{R}^n \times (0,\infty) \\ 0 & \text{if } (x,t) \in \mathbb{R}^n \times (-\infty, 0] - \{(0,0)\}. \end{cases} \quad (4.3.3)$$

It is easily seen that $E(\cdot,\cdot)$ is in $L^1_{loc}(\mathbb{R}^{n+1})$. We shall now show that the function $E$ is a fundamental solution of the heat operator. It is easy to verify that $E \in C^\infty(\mathbb{R}^{n+1} - \{0\})$. Moreover, $\left(\frac{\partial}{\partial t} - \Delta_x\right) E(x,t) = 0$ for $(x,t) \in (\mathbb{R}^{n+1} - \{0\})$. Suppose that $\varphi$ is in $\mathcal{D}(\mathbb{R}^{n+1})$ and $\epsilon > 0$, then we have

$$0 = \int_{\mathbb{R}^n} \int_\epsilon^\infty \varphi(x,t) \left(\frac{\partial}{\partial t} - \Delta_x\right) E(x,t)\, dx\, dt$$

$$= \int_{\mathbb{R}^n} \int_\epsilon^\infty \varphi(x,t) \frac{\partial}{\partial t} E(x,t)\, dx\, dt - \int_{\mathbb{R}^n} \int_\epsilon^\infty E(x,t) \Delta_x \varphi(x,t)\, dx\, dt.$$

Therefore

$$\int_{\mathbb{R}^n} \int_\epsilon^\infty \varphi(x,t) \frac{\partial}{\partial t} E(x,t)\, dx\, dt = \int_{\mathbb{R}^n} \int_\epsilon^\infty E(x,t) \Delta_x \varphi(x,t)\, dx\, dt. \quad (4.3.4)$$

Note that

$$\varphi \partial^{(0,\cdots,0,1)} E = \partial^{(0,\cdots,0,1)}(E\varphi) - E \partial^{(0,\cdots,0,1)} \varphi.$$

From this it follows that

$$\int_{\mathbb{R}^n} \int_\epsilon^\infty \frac{\partial}{\partial t}\left(E(x,t)\varphi(x,t)\right) dx\, dt - \int_{\mathbb{R}^n} \int_\epsilon^\infty E(x,t)\frac{\partial}{\partial t}\varphi(x,t)\, dx\, dt$$

$$= \int_{\mathbb{R}^n} \int_\epsilon^\infty E(x,t)\Delta_x\varphi(x,t)\, dx\, dt.$$

Equivalently, we have

$$\int_{\mathbb{R}^n} E(x,\epsilon)\varphi(x,\epsilon)\, dx$$

$$= -\int_{\mathbb{R}^n} \int_\epsilon^\infty E(x,t)\frac{\partial}{\partial t}\varphi(x,t)\, dx\, dt - \int_{\mathbb{R}^n} \int_\epsilon^\infty E(x,t)\Delta_x\varphi(x,t)\, dx\, dt.$$

Since $E$ is in $L^1_{loc}(\mathbb{R}^{n+1})$ therefore by the Lebesgue dominated convergence theorem we obtain

$$\lim_{\epsilon \to 0} \int_{\mathbb{R}^n} E(x, \epsilon)\varphi(x, \epsilon)\, dx = (D^{(0,\ldots,0,1)} - \Delta)E(\varphi).$$

It remains to prove that

$$\lim_{\epsilon \to 0} \int_{\mathbb{R}^n} E(x, \epsilon)\varphi(x, \epsilon)\, dx = \varphi(0,0).$$

We know that

$$\int_{\mathbb{R}^n} E(x, \epsilon)\varphi(x, \epsilon)\, dx = (4\pi\epsilon)^{-\frac{n}{2}} \int_{\mathbb{R}^n} \exp\left(-\frac{|x|^2}{4\epsilon}\right)\varphi(x, \epsilon)\, dx.$$

Putting $x = \sqrt{\epsilon}\omega$ we obtain

$$\int_{\mathbb{R}^n} E(x, \epsilon)\varphi(x, \epsilon)\, dx = (4\pi)^{-\frac{n}{2}} \int_{\mathbb{R}^n} \exp\left(-\frac{|\omega|^2}{4}\right)\varphi(\sqrt{\epsilon}\omega, \epsilon)\, d\omega.$$

Using again the Lebesgue dominated convergence theorem we get

$$\lim_{\epsilon \to 0} \int_{\mathbb{R}^n} E(x, \epsilon)\varphi(x, \epsilon)\, dx = \varphi(0,0).$$

Therefore $E$ is a fundamental solution of the heat operator.

## 4.4. The Hörmander inequalities

For solvability of differential equations in a bounded open set the known Hörmander inequality is very important. This section contains a proof of this inequality. To begin with, let us recall some notations. The norm and the inner product in $L^2(\Omega)$ will be denoted respectively by $\|\cdot\|$ and $\langle\cdot,\cdot\rangle$ or $\|\cdot\|_\Omega$ and $\langle\cdot,\cdot\rangle_\Omega$ if some ambiguity is possible. Let

$$P(\partial) = \sum_{|\nu|\leq m} a_\nu \partial^\nu$$

be a (non zero) linear differential operator with constant coefficients of order $m$.

THEOREM 4.4.1 (Hörmander). *For every bounded open set $\Omega$ in $\mathbb{R}^n$ there exists a constant $C > 0$ such that for every $\varphi \in \mathcal{D}(\Omega)$*

$$\|P(\partial)\varphi\| \geq C\|\varphi\|. \tag{4.4.1}$$

*As constant $C$, one can take $C = |P|_m K_{m,\Omega}$, where $|P|_m = \sup\{|a_j| : |j| = m\}$ and $K_{m,\Omega}$ depends only on $m$ and the diameter of $\Omega$.*

PROOF. Let $\varphi$ be in $\mathcal{D}(\Omega)$ and $A = \sup_\Omega |x|$. Of course, we have

$$\langle \partial^{e_j}(x_j\varphi), \varphi \rangle = \langle x_j \partial^{e_j}\varphi, \varphi \rangle + \langle \varphi, \varphi \rangle.$$

Integrating by parts we obtain

$$\langle \partial^{e_j}(x_j\varphi), \varphi \rangle = -\langle x_j\varphi, \partial^{e_j}\varphi \rangle.$$

Hence

$$\langle \varphi, \varphi \rangle = -\langle x_j \partial^{e_j}\varphi, \varphi \rangle - \langle x_j\varphi, \partial^{e_j}\varphi \rangle.$$

Therefore

$$\|\varphi\|^2 \leq 2A\|\varphi\|\,\|\partial^{e_j}\varphi\|.$$

Finally we have
$$\|\varphi\| \leq 2A\|\partial^{e_j}\varphi\|.$$
The general proof goes by induction on order of $P$. If $P(\partial)$ is a differential operator of order $m \geq 0$ then for $\varphi \in \mathcal{D}(\Omega)$ the following formula
$$P(\partial)(x_j\varphi) = x_j P(\partial)\varphi + P_j(\partial)\varphi \qquad (4.4.2)$$
holds. The operator $P_j(\partial)$ is zero if and only if $P(\partial)$ does not involve any differentiation with respect to $x_j$, and if non-zero, $P_j(\partial)$ is of order $< m$. By induction on $m$, we prove that for every $\varphi \in \mathcal{D}(\Omega)$
$$\|P_j(\partial)\varphi\| \leq 2mA\|P(\partial)\varphi\|. \qquad (4.4.3)$$
Note that this equality is true for $m = 1$. We shall now show that, if (4.4.3) holds, then the following inequality
$$\|P(\partial)(x_j\varphi)\| \leq (2m+1)A\|P(\partial)\varphi\| \qquad (4.4.4)$$
should be true. Indeed, by (4.4.2) and (4.4.3) we get (4.4.4). Assume now that (4.4.3) is true for all operators $P(\partial)$ of order $\leq m - 1$. Let the operator $P(\partial)$ be of order $m$ now. The expression $\langle P(\partial)(x_j\varphi), P_j(\partial)\varphi \rangle$ can be represented in two different ways. First, in accordance with the definition of $P_j(\partial)$ we have
$$P(\partial)(x_j\varphi) = x_j P(\partial)\varphi + P_j(\partial)\varphi.$$
From this we obtain
$$\langle P(\partial)(x_j\varphi), P_j(\partial)\varphi \rangle = \langle x_j P(\partial)\varphi, P_j(\partial)\varphi \rangle + \|P_j(\partial)\varphi\|^2. \qquad (4.4.5)$$
Second, applying integration by parts to $\langle P(\partial)(x_j\varphi), P_j(\partial)\varphi \rangle$, we get
$$\langle P(\partial)(x_j\varphi), P_j(\partial)\varphi \rangle = \langle P_j^*(\partial)(x_j\varphi), P^*(\partial)\varphi \rangle, \qquad (4.4.6)$$
where $P^*(\partial) = \sum_{|\nu| \leq m} \bar{a}_\nu(-1)^{|\nu|}\partial^\nu$. It is easily seen that
$$\|P(\partial)\varphi\| = \|P^*(\partial)\varphi\|. \qquad (4.4.7)$$
By (4.4.5) and (4.4.6) we obtain
$$\|P_j(\partial)\varphi\|^2 = \langle P_j^*(\partial)(x_j\varphi), P^*(\partial)\varphi \rangle - \langle x_j P(\partial)\varphi, P_j(\partial)\varphi \rangle. \qquad (4.4.8)$$
Since $P(\partial)$ is of order $m$, therefore $P_j(\partial)$ is of order $\leq m - 1$. Because (4.4.3) implies (4.4.4), therefore by (4.4.7) we get
$$\|P_j^*(\partial)(x_j\varphi)\| \leq (2m-1)A\|P_j(\partial)\varphi\|. \qquad (4.4.9)$$
By virtue of the Schwarz inequality we have
$$|\langle x_j P(\partial)\varphi, P_j(\partial)\varphi \rangle| \leq A\|P(\partial)\varphi\| \, \|P_j(\partial)\varphi\| \qquad (4.4.10)$$
and
$$|\langle P_j^*(\partial)(x_j\varphi), P^*(\partial)\varphi \rangle| \leq \|P_j^*(\partial)(x_j\varphi)\| \, \|P^*(\partial)\varphi\|. \qquad (4.4.11)$$
From (4.4.10) it follows the following inequality
$$-\langle x_j P(\partial)\varphi, P_j(\partial)\varphi \rangle \leq A\|P(\partial)\varphi\| \, \|P_j(\partial)\varphi\|. \qquad (4.4.12)$$
In view of (4.4.8), (4.4.12), (4.4.10) and (4.4.11) we get
$$\|P_j(\partial)\varphi\|^2 \leq \|P_j^*(\partial)(x_j\varphi)\| \, \|P^*(\partial)\varphi\| + A\|P(\partial)\varphi\| \, \|P_j(\partial)\varphi\|$$
Taking into account (4.4.7) and (4.4.9) we obtain
$$\|P_j(\partial)\varphi\|^2 \leq (2m-1)A\|P_j(\partial)\varphi\| \, \|P(\partial)\varphi\| + A\|P(\partial)\varphi\| \, \|P_j(\partial)\varphi\|.$$

Finally, we have

$$\|P_j(\partial)\varphi\| \leq 2mA\|P(\partial)\varphi\|.$$

Thus, the proof of (4.4.3) is complete.

We are now in a position to finish the proof. Let $P(\partial)$ be an operator of order 1. It can be written as follows

$$P(\partial) = a_1\partial^{e_1} + \cdots + a_n\partial^{e_n} + a_0.$$

Note that $P(\partial)(x_j\varphi) = x_jP(\partial)\varphi + a_j\varphi$. Therefore $P_j(\partial) = a_j$. Let us choose $j$ so that $|a_j| = \max_{\nu=1,\ldots,n} |a_\nu|$. Of course, $|a_j| = |P_j|_0 \geq |P|_1$ and $|P_j|_0\|\varphi\| \leq 2A\|P(\partial)\varphi\|$. In general, one can show that if $P(\partial)$ is an operator of order $m \geq 1$, then there exists $j \in \{1,\ldots,n\}$ so that $P_j(\partial)$ is of order $m-1$ and $|P_j|_{m-1} \geq |P|_m$. Taking into account this remark and (4.4.3) immediately we obtain (4.4.1) by induction on $m$.                                                                              $\square$

Equality (4.4.1) is called Hörmander's inequality.

EXAMPLE 4.4.1. Let $\Omega$ be an open bounded set. Note that

$$\langle -\Delta\varphi, \varphi\rangle = \sum_{\nu=1}^{n}\langle \partial^{e_\nu}\varphi, \partial^{e_\nu}\varphi\rangle.$$

Applying (4.4.1) for the operator $\partial^{e_\nu}$, $\nu = 1,\ldots n$ we have

$$\langle -\Delta\varphi, \varphi\rangle \geq nC_\Omega'\langle\varphi, \varphi\rangle.$$

Taking $C_\Omega = nC_\Omega'$ we obtain the Poincaré inequality

$$\langle -\Delta\varphi, \varphi\rangle \geq C_\Omega\|\varphi\|^2 \tag{4.4.13}$$

for $\varphi \in \mathcal{D}(\Omega)$.

In the sequel we shall need a new inequality with a weight. Similarly as above, let $P(\partial)$ be a linear differential operator with constant coefficients.

THEOREM 4.4.2 (Hörmander). *For every bounded open set $\Omega$ in $\mathbb{R}^n$ there exists a constant $M > 0$ such that for all $\eta \in \mathbb{R}$ and $\varphi \in \mathcal{D}(\Omega)$*

$$\int\limits_{\Omega} e^{\eta x_j}|P(\partial)\varphi(x)|^2\, dx \geq M \int\limits_{\Omega} e^{\eta x_j}|\varphi(x)|^2\, dx, \tag{4.4.14}$$

*where $x = (x_1,\ldots,x_j,\ldots,x_n)$.*

PROOF. Define an operator $Q(\partial)$ as follows

$$Q(\partial) := e^{\frac{\eta}{2}x_j}P(\partial)e^{-\frac{\eta}{2}x_j}.$$

In the first place we shall show that $Q(\partial)$ is a constant coefficient operator with the same terms of higher degree as $P(\partial)$. Note that we can restrict our considerations to ordinary differential operators. Let

$$P\left(\frac{d}{dx}\right) = a_m\frac{d^m}{dx^m} + a_{m-1}\frac{d^{m-1}}{dx^{m-1}} + \cdots + a_1\frac{d}{dx} + a_0, \quad a_i \in \mathbb{C}, \quad a_m \neq 0.$$

Compute

$$\left(Q(\frac{d}{dx})\psi\right)(x) = \left(e^{\frac{\eta}{2}x}P(\frac{d}{dx})e^{-\frac{\eta}{2}x}\right)\psi(x) = \sum_{\nu=0}^{m}a_\nu\left(e^{\frac{\eta}{2}x}\frac{d^\nu}{dx^\nu}e^{-\frac{\eta}{2}x}\right)\psi(x).$$

Note that

$$\frac{d^\nu}{dx^\nu}\left(e^{-\frac{\eta}{2}x}\psi(x)\right) = e^{-\frac{\eta}{2}x}\frac{d^\nu}{dx^\nu}\psi(x) + \sum_{\mu=1}^{\nu}\binom{\nu}{\mu}\left(-\frac{\eta}{2}\right)^\mu e^{-\frac{\eta}{2}x}\frac{d^{\nu-\mu}}{dx^{\nu-\mu}}\psi(x).$$

Hence

$$\left(e^{\frac{\eta}{2}x}\frac{d^\nu}{dx^\nu}e^{-\frac{\eta}{2}x}\right)\psi(x) = \frac{d^\nu}{dx^\nu}\psi(x) + \sum_{\mu=1}^{\nu}\binom{\nu}{\mu}\left(-\frac{\eta}{2}\right)^\mu\frac{d^{\nu-\mu}}{dx^{\nu-\mu}}\psi(x).$$

From this, it is easily seen that $Q(\partial)$ is indeed a constant coefficient operator with the same terms of higher degree as $P(\partial)$. We shall now apply (4.4.1) to the function $\psi = e^{\frac{\eta}{2}x}\varphi(x)$ with respect to $Q(\partial)$. Then we have

$$\int_{\mathbb{R}^n}\left\|e^{\frac{\eta}{2}x_j}P(\partial)\varphi(x)\right\|^2 dx \geq C^2\int_{\mathbb{R}^n}\left|e^{\frac{\eta}{2}x_j}\varphi(x)\right|^2 dx.$$

Finally, if we take $M = C^2$ we obtain (4.4.14). $\qquad\square$

COROLLARY 4.4.1. *If $\varphi \in \mathcal{D}(\mathbb{R}^n)$, $P \neq 0$ and $P(\partial)\varphi(x) = 0$ in $\Pi_j^+ = \{x : x_j \geq 0\}$, then $\varphi(x) = 0$ in $\Pi_j^+$, too.*

PROOF. Let $\Omega$ be an open set containing supp $\varphi$. Suppose, contrary to our claim, that $\varphi$ does not vanish in $\Pi_j^+$, then one can choose $\eta > 0$ so that the expression on the right side of (4.4.14) will be arbitrarily big, but the expression on the left side of (4.4.14) will be bounded for $\eta > 0$. This gives a contradiction. Thus, the proof is finished. $\qquad\square$

We are now in a position to prove a more general

THEOREM 4.4.3. *Let $\varphi \in \mathcal{D}(\mathbb{R}^n)$, or more generally $\varphi \in L^2(\mathbb{R}^n)$, $\varphi$ has a compact support. If $P \neq 0$ and $P(D)\varphi = 0$ in $\mathbb{R}^n - \overline{B_r(0)}$ (if $\varphi$ is not smooth, then $P(D)\varphi = 0$ in the distributional sense), then supp $\varphi$ is in $\overline{B_r(0)}$.*

PROOF. We shall first prove the smooth case. Note that the ball $\overline{B_r(0)}$ is the intersection of half spaces

$$\Pi_{rh} = \{x : x_1 h_1 + \cdots + x_n h_n \geq r, |h| = 1\}.$$

Suppose that $P(\partial)\varphi$ is equal to zero in $\mathbb{R}^n - \overline{B_r(0)}$, then by using translations and rotations, we infer in view of Corollary 4.4.1, that supp $\varphi \subset \overline{B_r(0)}$. To treat the non-smooth case, take

$$\varphi_\epsilon = \varphi * \psi_\epsilon,$$

where $\psi_\epsilon(x) = \epsilon^{-n}\psi\left(\frac{x}{\epsilon}\right)$, $\psi \in \mathcal{D}(B_1(0))$ and $\int \psi = 1$. Then $\varphi_\epsilon \in \mathcal{D}(\mathbb{R}^n)$ and supp $\varphi_\epsilon \subset B_\epsilon(\text{supp } \varphi)$. Note that

$$P(\partial)\varphi_\epsilon(x) = \int_{\mathbb{R}^n} P(\partial_x)\psi_\epsilon(x-y)\varphi(y)\,dy = \int_{\mathbb{R}^n} P(-\partial_y)\psi_\epsilon(x-y)\varphi(y)\,dy.$$

Let $x \notin \overline{B_{\epsilon+r}(0)}$, then supp $\psi_\epsilon(x-\cdot) \subset \mathbb{R}^n - B_r(0)$. By the assumption we get $P(\partial)\varphi_\epsilon(x) = 0$. This reduces the problem to the smooth case. By Lemma 1.2.2, $\varphi_\epsilon \to \varphi$ in $L^2(\mathbb{R}^n)$ as $\epsilon \to 0$. Thus, the proof is finished. $\qquad\square$

## 4.5. $L^2$-solvability

Let $P(D)$ be a linear differential operator in the distributional sense with constant coefficients. Using the Hörmander inequality we prove the following

THEOREM 4.5.1. *If $\Omega$ is an open bounded set in $\mathbb{R}^n$, then for every $g \in L^2(\Omega)$ there exists $u \in L^2(\Omega)$ such that $P(D)u = g$ in $\mathcal{D}'(\Omega)$.*

PROOF. Since $\|P^*(\partial)\varphi\| = \|P(\partial)\varphi\|$ (see (4.4.7)) therefore by (4.4.1) we have $\|P^*(\partial)\varphi\| \geq C\|\varphi\|$ for $\varphi \in \mathcal{D}(\Omega)$. Let

$$\mathcal{X} = \{P^*(\partial)\varphi : \varphi \in \mathcal{D}(\Omega)\}.$$

Consider the following mapping

$$\mathcal{X} \ni P^*(\partial)\varphi \to \langle g, \varphi \rangle. \tag{4.5.1}$$

Note that

$$|\langle g, \varphi \rangle| \leq \|g\| \, \|\varphi\| \leq C^{-1}\|g\| \, \|P^*(\partial)\varphi\|.$$

This means that the mapping (4.5.1) is continuous on $\mathcal{X}$ with respect to the norm $\|\cdot\|$. It can be extended on $\overline{\mathcal{X}}$, the closure of $\mathcal{X}$ in $L^2(\Omega)$. Then the Riesz theorem gives the existence of $u \in \overline{\mathcal{X}}$, so that $\langle g, \varphi \rangle = \langle u, P^*(\partial)\varphi \rangle$. This means that $P(D)u = g$ in $\mathcal{D}'(\Omega)$. Thus, the proof of the theorem is finished. $\square$

To solve the differential equations in $\mathbb{R}^n$ we need the following approximation

THEOREM 4.5.2. *Let $0 < r < r' < R$. If $v \in L^2(B_{r'}(0))$ and $v$ satisfies $P(D)v = 0$ on $B_{r'}(0)$, then there exists a sequence $(v_\nu)$, $v_\nu \in L^2(B_R(0))$ so that $P(D)v_\nu = 0$ on $B_R(0)$ and $v_\nu$ tends to $v$ in $L^2(B_r(0))$ as $\nu \to \infty$.*

PROOF. Let

$$A = \{\alpha : \alpha \in L^2(B_R(0)), \ P(D)\alpha = 0 \text{ in } \mathcal{D}'(B_R(0))\}$$

and let $A^\perp$ denote the set of all $g \in L^2(B_r(0))$ which are orthogonal to every $\alpha \in A$ in the sense of $L^2(B_r(0))$. We shall first show that if $g \in A^\perp$, then

$$|\langle \varphi, g \rangle_{B_r}| \leq C\|P(\partial)\varphi\|_{B_R} \tag{4.5.2}$$

for $\varphi \in \mathcal{D}(\mathbb{R}^n)$, where $\|\cdot\|_{B_r} := \|\cdot\|_{L^2(B_r(0))}$. Let us consider two cases, namely $P(\partial)\varphi = 0$ and $P(\partial)\varphi \neq 0$. Obviously, we have $\langle \varphi, g \rangle_{B_r} = 0$, when $P(\partial)\varphi = 0$. We shall now consider the case, when $P(\partial)\varphi \neq 0$. Let $\overline{\mathcal{X}}$ denote the closure of $\mathcal{X} = \{P^*(\partial)\Theta : \Theta \in \mathcal{D}(B_R(0))\}$ in $L^2(B_R(0))$. By Theorem 4.5.1 there exists $\psi$ in $\overline{\mathcal{X}}$ such that

$$\langle \psi, P^*(\partial)\Theta \rangle_{B_R} = \langle P(\partial)\varphi, \Theta \rangle_{B_R}$$

for $\Theta \in \mathcal{D}(B_R(0))$. Hence, by (4.4.1) we have

$$|\langle \psi, P^*(\partial)\Theta \rangle_{B_R}| = |\langle P(\partial)\varphi, \Theta \rangle_{B_R}| \leq \|P(\partial)\varphi\|_{B_R}\|\Theta\|_{B_R}$$
$$\leq C_1\|P(\partial)\varphi\|_{B_R}\|P^*(\partial)\Theta\|_{B_R}.$$

Therefore

$$|\langle \psi, P^*(\partial)\Theta \rangle_{B_R}| \leq C_1\|P(\partial)\varphi\|_{B_R}\|P^*(\partial)\Theta\|_{B_R}.$$

It is easily seen that

$$|\langle \psi, \chi \rangle_{B_R}| \leq C_1\|P(\partial)\varphi\|_{B_R}\|\chi\|_{B_R}$$

if $\chi \in \overline{\mathcal{X}}$. Finally we have

$$\|\psi\|_{B_R} \leq C_1\|P(\partial)\varphi\|_{B_R}. \tag{4.5.3}$$

Of course,

$$\langle \varphi, g \rangle_{B_r} = \langle \varphi - \psi, g \rangle_{B_r} + \langle \psi, g \rangle_{B_r}.$$

Since $\psi - \varphi$ is in $A$, therefore

$$\langle \varphi, g \rangle_{B_r} = \langle \psi, g \rangle_{B_r}.$$

Hence by (4.5.3) we get

$$|\langle \varphi, g \rangle_{B_r}| \leq C\|P(\partial)\varphi\|_{B_R}$$

for $\varphi \in \mathcal{D}(\mathbb{R}^n)$, where $C = C_1\|g\|_{B_r}$. Thus, inequality (4.5.2) is proved. Let

$$\mathcal{Y} = \{P(\partial)\varphi : \varphi \in \mathcal{D}(\mathbb{R}^n)\}.$$

By (4.5.2) the mapping $\mathcal{Y} \ni P(\partial)\varphi \to \langle \varphi, g \rangle_{B_r}$ is a continuous linear form on $\mathcal{Y}$ with respect to the norm $\| \cdot \|_{B_R}$. This mapping can be extended by continuity from $\mathcal{Y}$ onto $\overline{\mathcal{Y}}$ (the closure of $\mathcal{Y}$ in $L^2(B_R(0))$. Then the Riesz theorem gives the existence of $w$ in $\overline{\mathcal{Y}}$ so that

$$\langle \varphi, g \rangle_{B_r} = \langle w, P(\partial)\varphi \rangle_{B_R} \tag{4.5.4}$$

for $\varphi \in \mathcal{D}(\mathbb{R}^n)$. Set

$$\tilde{w}(x) := \begin{cases} w(x) & \text{for } x \in B_R(0), \\ 0 & \text{for } x \in \mathbb{R}^n - B_R(0) \end{cases}$$

and

$$\tilde{g}(x) := \begin{cases} g(x) & \text{for } x \in B_r(0), \\ 0 & \text{for } x \in \mathbb{R}^n - B_r(0). \end{cases}$$

Hence, equality (4.5.4) can be written as follows

$$\langle \varphi, \tilde{g} \rangle_{\mathbb{R}^n} = \langle \tilde{w}, P(\partial)\varphi \rangle_{\mathbb{R}^n}. \tag{4.5.5}$$

This means that $\tilde{g} = P^*(D)\tilde{w}$ in $\mathcal{D}'(\mathbb{R}^n)$. Since supp $\tilde{w}$ is a compact set and supp $\tilde{g} \subset B_r(0)$, therefore by Theorem 4.4.3 supp $\tilde{w} \subset B_r(0)$. This implies that

$$w(x) = 0 \quad \text{if} \quad x \in B_R(0) - B_r(0). \tag{4.5.6}$$

We begin now the last step of the proof. Suppose that, $v \in L^2(\mathbb{R}^n)$, supp $v \subset B_{r'}(0)$ and $P(D)v = 0$ in $\mathcal{D}'(B_{r'}(0))$. Then $P(\partial)v_\epsilon(x) = 0$ for $x \in B_r(0)$, $\epsilon < \epsilon_0 < \min(r' - r, R - r')$, where $v_\epsilon = \psi_\epsilon * v$ (see the proof of Theorem 4.4.3). Moreover, $v_\epsilon \to v$ as $\epsilon \to 0$ in $L^2(B_{r'+\epsilon_0}(0))$ and $v_\epsilon \in \mathcal{D}(B_{r'+\epsilon_0}(0))$. Taking into account (4.5.6) and putting $v_\epsilon$ in (4.5.4) in the place of $\varphi$ we obtain

$$\langle v_\epsilon, g \rangle_{B_r} = \langle w, P(\partial)v_\epsilon \rangle_{B_r} = 0. \tag{4.5.7}$$

Hence we have

$$\lim_{\epsilon \to 0} \langle v_\epsilon, g \rangle_{B_r} = \langle v, g \rangle_{B_r} = 0.$$

Since $g$ is an arbitrary element of $A^\perp$, therefore $v \in (A^\perp)^\perp = \overline{A}$, where $\overline{A}$ denotes the closure of $A$ in $L^2(B_r(0))$. From this it follows that there exists a sequence $(v_\nu)$, $v_\nu \in A$ such that $v_\nu$ tends to $v$ as $\nu \to \infty$ in $L^2(B_r(0))$. Thus, the proof is complete. $\square$

The next theorem gives us the existence of solutions of differential equations with constant coefficients in $\mathbb{R}^n$.

THEOREM 4.5.3. *Let $P(D)$ be a (non-zero) constant coefficient linear differential operator in distributional sense on $\mathbb{R}^n$. Then for every $g \in L^2_{loc}(\mathbb{R}^n)$ there exists $u \in L^2_{loc}(\mathbb{R}^n)$ such that $P(D)u = g$.*

PROOF. By Theorem 4.5.1 there exists $u_1 \in L^2(B_2(0))$ such that $P(D)u_1 = g$. Let $w$ be an arbitrary function in $L^2(B_3(0))$ fulfilling $P(D)w = g$. In view of Theorem 4.5.2 there exists $v$ in $L^2(B_3(0))$ such that $P(D)v = 0$ on $B_3(0)$ and

$$\|u_1 - w - v\|_{B_1} < \frac{1}{2}.$$

Put $u_2 = w + v$, then $P(D)u_2 = g$ on $B_3(0)$ and

$$\|u_2 - u_1\|_{B_1} < \frac{1}{2}.$$

Inductively, assuming that $u_\nu$ has been chosen in $L^2(B_{\nu+1}(0))$ so that $P(D)u_\nu = g$ on $B_{\nu+1}(0)$, one can choose $u_{\nu+1}$ in $L^2(B_{\nu+2}(0))$ in the following way. Let $w$ be in $L^2(B_{\nu+2}(0))$ and $P(D)w = g$ on $B_{\nu+2}(0)$. On $B_{\nu+1}(0)$ one has $P(D)(u_\nu - w) = 0$. By virtue of Theorem 4.5.2 there exists $v \in L^2(B_{\nu+2}(0))$ such that $P(D)v = 0$ on $B_{\nu+2}(0)$ and

$$\|u_\nu - w - v\|_{B_\nu} < \frac{1}{2^\nu}.$$

Similarly as above, set $u_{\nu+1} = w + v$. Then $P(D)u_{\nu+1} = g$ on $B_{\nu+2}(0)$ and

$$\|u_{\nu+1} - u_\nu\|_{B_\nu} < \frac{1}{2^\nu}.$$

Of course, the sequence $(u_\nu)$ converges to some $u$ in $L^2_{loc}(\mathbb{R}^n)$. This implies that $P(D)u_\nu$ converges to $P(D)u$ in $\mathcal{D}'(\mathbb{R}^n)$. Since $P(D)u_\nu = g$ in $\mathcal{D}'(B_\nu(0))$, therefore $P(D)u = g$ in $\mathcal{D}'(\mathbb{R}^n)$. This finishes the proof.  $\square$

We are able to prove the main theorem of this section.

THEOREM 4.5.4 (Malgrange-Ehrenpreis). *Every nonzero linear differential operator $P(D)$ with constant coefficients on $\mathbb{R}^n$ has a fundamental solution.*

PROOF. Let

$$H(x) = H(x_1) \cdots H(x_n), \quad x = (x_1, \ldots, x_n),$$

($H$ is the Heaviside step function), then $D^{(1,\ldots,1)}H = \delta$. By Theorem 4.5.3 there exists $u \in L^2_{loc}(\mathbb{R}^n)$, $P(D)u = H$. Take $E = D^{(1,\ldots,1)}u$. Then we have

$$P(D)E = P(D)D^{(1,\ldots,1)}u = D^{(1,\ldots,1)}P(D)u = D^{(1,\ldots,1)}H = \delta.$$

This ends the proof.  $\square$

## 4.6. Regularity properties of differential operators

Let $\Lambda$ be a compact support distribution and $\Theta \in C^\infty(\mathbb{R}^n)$. Put

$$\Lambda_\Theta(x) := \Lambda(\Theta(x - \cdot)).$$

THEOREM 4.6.1. *The function $\Lambda_\Theta$ is in $C^\infty(\mathbb{R}^n)$.*

PROOF. It is easily seen that $\Theta(x - \cdot)$ tends to $\Theta(x_0 - \cdot)$ as $x \to x_0$ in $C^\infty(\mathbb{R}^n)$. The rest of the proof runs as in the proof of Theorem 3.1.1.  $\square$

DEFINITION 4.6.1. The function $\Lambda_\Theta$ is called the regularization of $\Lambda$ by means of the function $\Theta$.

THEOREM 4.6.2. *If $\Lambda$ is a compact support distribution and $\Theta \in C^\infty(\mathbb{R}^n)$, then the convolution product $\Lambda * \Theta$ is in $C^\infty(\mathbb{R}^n)$ and $\Lambda * \Theta = \Lambda_\Theta$.*

PROOF. To this end we have to show that

$$(\Lambda_y \otimes \Theta_x)\varphi(x+y) = \int_{\mathbb{R}^n} \Lambda_\Theta(\omega)\varphi(\omega)\,d\omega$$

for $\varphi \in \mathcal{D}(\mathbb{R}^n)$. In accordance with the definition of $\Lambda \otimes \Theta$ we have

$$(\Lambda_y \otimes \Theta_x)\varphi(x+y) = \Lambda_y \left( \int_{\mathbb{R}^n} \Theta(x)\varphi(x+y)\,dx \right) = \Lambda_y \left( \int_{\mathbb{R}^n} \varphi(\omega)\Theta(\omega-y)\,d\omega \right).$$

Set by definition

$$\sigma_h(y) := \sum_{\nu \in \mathbb{N}_h^n} \varphi(h\nu)\Theta(h\nu - y)h^n, \tag{4.6.1}$$

where $\mathbb{N}_h^n = \{\nu : h\nu \subset \operatorname{supp}\varphi\}$ (compare the proof of Theorem 3.1.4). One can show that $\sigma_h(y)$ tends to $\int_{\mathbb{R}^n} \varphi(\omega)\Theta(\omega-y)\,d\omega$ as $h \to 0$ in the sense of $C^\infty$-convergence. From this it follows that $\Lambda(\sigma_h)$ tends to $\Lambda_y\left( \int_{\mathbb{R}^n} \varphi(\omega)\Theta(\omega-y)\,d\omega \right)$. By (4.6.1) we have

$$\Lambda(\sigma_h) = \sum_{\nu \in \mathbb{N}_h^n} \varphi(h\nu)\Lambda_y(\Theta(h\nu - y))h^n. \tag{4.6.2}$$

The expression on the right side of (4.6.2) is a Riemann sum of the integral $\int_{\mathbb{R}^n} \varphi(\omega)\Lambda_\Theta(\omega)\,d\omega$. Therefore $\Lambda(\sigma_h)$ tends to $\int_{\mathbb{R}^n} \varphi(\omega)\Lambda_\Theta(\omega)\,d\omega$ as $h \to 0$. From the above consideration it follows that

$$(\Lambda_y \otimes \Theta_x)\varphi(x+y) = \int_{\mathbb{R}^n} \varphi(\omega)\Lambda_\Theta(\omega)\,d\omega.$$

This statement finishes the proof. $\qquad\square$

DEFINITION 4.6.2. The singular support of a distribution $\Lambda$ is defined to be the complement of the largest open set on which $\Lambda$ is a $C^\infty$-function.

The singular support of $\Lambda$ will be denoted by $\operatorname{sing\,supp}\Lambda$.

THEOREM 4.6.3. Let $\Lambda$ be in $\mathcal{D}'(\mathbb{R}^n)$. If $\Lambda$ is a $C^\infty$-function on $\mathbb{R}^n - \{0\}$ and $S$ is a compact support distribution, then $\operatorname{sing\,supp}\Lambda * S \subset \operatorname{supp}S$.

PROOF. We only need to show that if $x$ does not belong to $\operatorname{supp}S$, then there exists a neighbourhood $U_x$ of $x$ such that the restriction $\Lambda * S|_{U_x}$ of $\Lambda * S$ belongs to $C^\infty(U_x)$. To this end assume that $x \notin \operatorname{supp}S$, therefore there exists $\epsilon > 0$ such that $B_\epsilon(x) \cap \operatorname{supp}S = \emptyset$. Take $\psi \in \mathcal{D}(B_{\frac{\epsilon}{2}}(0))$, $\psi \geq 0$ and $\psi(x) = 1$ on $\overline{B_{\frac{\epsilon}{4}}(0)}$, then by Theorem 3.7.2 it follows that

$$\operatorname{supp}((\psi\Lambda) * S) \subset \overline{\operatorname{supp}S + B_{\frac{\epsilon}{2}}(0)}.$$

This implies that $((\psi\Lambda) * S)|_{B_{\frac{\epsilon}{4}}(x)} = 0$. Hence we have

$$(\Lambda * S)|_{B_{\frac{\epsilon}{4}}(x)} = ((1-\psi)\Lambda * S)|_{B_{\frac{\epsilon}{4}}(x)}.$$

Obviously $(1-\psi)\Lambda \in C^\infty(\mathbb{R}^n)$. By Theorem 4.6.2

$$(\Lambda * S)|_{B_{\frac{\epsilon}{4}}(x)} \in C^\infty(B_{\frac{\epsilon}{4}}(x)).$$

Since $x$ was arbitrarily chosen from the complement of $\operatorname{supp}S$, therefore $\Lambda * S$ is a smooth function on the outside $\operatorname{supp}S$. Thus, the proof is finished. $\qquad\square$

DEFINITION 4.6.3. A differential operator $P(D) = \sum_{|\alpha| \leq m} a_\alpha D^\alpha$, $a_\alpha \in \mathbb{C}$ is said to be hypoelliptic if for any $\Lambda \in \mathcal{D}'(\mathbb{R}^n)$, sing supp $\Lambda \subset$ sing supp $P(D)\Lambda$.

In other words, $P(D)$ is hypoelliptic if and only if for any open set $\Omega \subset \mathbb{R}^n$ and $\Lambda \in \mathcal{D}'(\Omega)$ the following implication

$$P(D)\Lambda \in C^\infty(\Omega) \Rightarrow \Lambda \in C^\infty(\Omega)$$

holds.

THEOREM 4.6.4. *Let $P(D)$ be a partial differential operator with constant coefficients. Then the following conditions are equivalent:*

   (i) *$P(D)$ is hypoelliptic,*
   (ii) *every fundamental solution of $P(D)$ is $C^\infty$ in $\mathbb{R}^n - \{0\}$,*
   (iii) *at least one fundamental solution of $P(D)$ is $C^\infty$ in $\mathbb{R}^n - \{0\}$.*

PROOF. A fundamental solution $E$ of the operator $P(D)$ satisfies the equation $P(D)E = \delta$. Of course, if $P(D)$ is hypoelliptic, then every fundamental solution of $P(D)$ is a $C^\infty$-function in $\mathbb{R}^n - \{0\}$, therefore (i) implies (ii). The implication (ii)$\Rightarrow$(iii) is completely trivial. We only need to show that (iii) implies (i). Let $E$ be a fundamental solution of $P(D)$. Suppose that $E$ is $C^\infty$ in $\mathbb{R}^n - \{0\}$. Take $u \in \mathcal{D}'(\mathbb{R}^n)$ and assume that $P(D)u \in C^\infty(\Omega)$, where $\Omega \subset \mathbb{R}^n$ is open. Fix $x$ in $\Omega$, then $B_\epsilon(x) \subset \Omega$ for some $\epsilon > 0$. Take a function $\psi$ in $\mathcal{D}(B_\epsilon(x))$ such that $\psi(x) \geq 0$, $\psi(x) = 1$ for $x \in B_{\frac{\epsilon}{2}}(x)$. Note that

$$P(D)(\psi u) = \psi P(D)u + v,$$

where $v = 0$ on $B_{\frac{\epsilon}{2}}(x)$ and on outside $B_\epsilon(x)$. Of course,

$$E * (P(D)(\psi u)) = E * (\psi P(D)u) + E * v.$$

Since $\psi P(D)u$ is in $\mathcal{D}(B_\epsilon(x))$, therefore by Theorem 3.1.2 and Theorem 3.6.1, $E * (\psi P(D)u)$ belongs to $C^\infty(\mathbb{R}^n)$. Taking into account that supp $v$ is on outside $B_{\frac{\epsilon}{2}}(x)$, by Theorem 4.6.3 we infer that sing supp$(E * v) \cap B_{\frac{\epsilon}{2}}(x) = \emptyset$. Therefore $E * P(D)(\psi u)$ is a $C^\infty$-function on $B_{\frac{\epsilon}{2}}(x)$. Note that

$$E * P(D)(\psi u) = P(D)E * (\psi u) = \delta * (\psi u) = \psi u.$$

By the above observation $\psi u$ is a $C^\infty$-function on $B_{\frac{\epsilon}{2}}(x)$. Therefore $u$ is smooth on $B_{\frac{\epsilon}{2}}(x)$. This statement finishes the proof.                      $\square$

REMARK 4.6.1. The Laplace operator and the heat operator are hypoelliptic but the wave operator is not hypoelliptic. For proof see Sections 4.2 and 4.3. Compare also Example 1.9.1.

## CHAPTER 5

# PARTICULAR TYPES OF DISTRIBUTIONS AND CAUCHY TRANSFORMS

We shall present in this chapter new subspaces $\mathcal{D}'_{L^p}$, $1 \leq p < \infty$ of the space $\mathcal{D}'$, which are natural generalization of $L^p$-spaces. The distributions belonging to $\mathcal{D}'_{L^p}$ are important for applications and are simply represented by analytic functions.

## 5.1. Integrable distributions

DEFINITION 5.1.1. We say that $\varphi$ is in $\overset{\circ}{\mathcal{B}}$ if $\varphi$ is smooth and $\partial^\alpha \varphi$ belongs to $C_0(\mathbb{R}^n)$ for each $\alpha \in \mathbb{N}^n$.

DEFINITION 5.1.2. Let $\varphi_\nu$ be in $\overset{\circ}{\mathcal{B}}$, $\nu = 1, 2, \ldots$. We say that the sequence $(\varphi_\nu)$ converges to zero in $\overset{\circ}{\mathcal{B}}$ if for each nonnegative integer $m$, the sequence $(\|\varphi_\nu\|_{m,\mathbb{R}^n})$ tends to zero as $\nu \to \infty$.

THEOREM 5.1.1. *The space $\mathcal{D}$ is dense in $\overset{\circ}{\mathcal{B}}$.*

PROOF. Let $\varphi$ be in $\overset{\circ}{\mathcal{B}}$. Put $\varphi_\nu = g_\nu \varphi$, where $g_\nu$ is taken as in Lemma 1.12.1. It is evident that $\|\varphi_\nu - \varphi\|_{m,\mathbb{R}^n}$ tends to zero as $\nu \to \infty$ for $m = 0, 1, 2, \ldots$. Since the functions $\varphi_\nu$ are in $\mathcal{D}$, $\mathcal{D}$ is dense in $\overset{\circ}{\mathcal{B}}$. This finishes the proof. $\square$

The following theorem gives us a representation formula for the linear continuous forms on $\overset{\circ}{\mathcal{B}}$.

THEOREM 5.1.2. *For an arbitrary continuous linear form $\Lambda$ on $\overset{\circ}{\mathcal{B}}$ there exist complex regular Borel measures $\mu_\alpha$ on $\mathbb{R}^n$ such that*

$$\Lambda(\varphi) = \sum_{|\alpha| \leq m} (-1)^{|\alpha|} \int_{\mathbb{R}^n} \partial^\alpha \varphi \, d\mu_\alpha \quad for \quad \varphi \in \overset{\circ}{\mathcal{B}}. \tag{5.1.1}$$

Since the proof of this theorem is the same as the proof of Theorem 2.7.3, we will not give it here.

We shall now consider a larger vector space than $\overset{\circ}{\mathcal{B}}$.

DEFINITION 5.1.3. We say that a function $\varphi$ belongs to $\mathcal{D}_{L^\infty}$ if it is smooth and all of its derivatives $\partial^\alpha \varphi$, $\alpha \in \mathbb{N}^n$ are bounded on $\mathbb{R}^n$.

On the space $\mathcal{D}_{L^\infty}$ ([24]) we introduce a convergence in the following way.

DEFINITION 5.1.4. A sequence $(\varphi_\nu)$, $\varphi_\nu \in \mathcal{D}_{L^\infty}$ is said to be $\beta$-convergent to zero in $\mathcal{D}_{L^\infty}$ if it satisfies the following two conditions:

(i) there exists a positive real number $M_\alpha$ such that $\|\partial^\alpha \varphi_\nu\|_{L^\infty} \leq M_\alpha$ for $\alpha \in \mathbb{N}^n$, $\nu = 1, 2, \ldots$,

(ii) the sequence $(\partial^\alpha \varphi_\nu)$ uniformly converges to zero on each compact set $K \subset \mathbb{R}^n$.

THEOREM 5.1.3. *The space $\mathcal{D}$ is dense in $\mathcal{D}_{L^\infty}$ with respect to $\beta$-convergence.*

PROOF. Note that $\mathbb{R}^n = \bigcup_{\nu=1}^\infty B_\nu(0)$. Of course, $\overline{B_\nu(0)}$, $\nu = 1, 2, \ldots$ is compact. Let $g_\nu$ be as in Lemma 1.12.1. Then there exist constants $K_\alpha$ such that $\|\partial^\alpha g_\nu\|_{L^\infty} \leq K_\alpha$ for $\nu = 1, 2, \ldots$. For $\varphi$ in $\mathcal{D}_{L^\infty}$ take $\varphi_\nu = g_\nu \varphi$. Note that $\varphi_\nu$ belongs to $\mathcal{D}$ and

$$\partial^\alpha(g_\nu \varphi) = g_\nu \partial^\alpha \varphi + \sum_{0 < \beta \leq \alpha} \binom{\alpha}{\beta} \partial^\beta g_\nu \partial^{\alpha-\beta} \varphi.$$

It is easily seen that $\partial^\beta g_\nu(x) = 0$ for $x \in B_{\nu-1}(0)$ and $|\beta| > 0$. Hence we get

$$\frac{\partial^{|\alpha|}}{\partial x^\alpha}(g_\nu \varphi)(x) = \frac{\partial^{|\alpha|}}{\partial x^\alpha} \varphi(x)$$

for $x \in B_{\nu-1}(0)$. Therefore

$$\|\partial^\alpha(g_\nu \varphi) - \partial^\alpha \varphi\|_{L^\infty(B_k(0))} = 0$$

if $\nu \geq k + 1$. This implies (ii). On the other hand we have

$$\|\partial^\alpha(g_\nu \varphi)\|_{L^\infty} \leq \sum_{0 \leq \beta \leq \alpha} \binom{\alpha}{\beta} \|\partial^\beta g_\nu\|_{L^\infty} \|\partial^{\alpha-\beta} \varphi\|_{L^\infty}$$

$$\leq \sum_{0 \leq \beta \leq \alpha} \binom{\alpha}{\beta} K_\beta \|\partial^{\alpha-\beta} \varphi\|_{L^\infty} =: M_\alpha.$$

Thus, the sequence $(\varphi_\nu)$ satisfies (i). This statement finishes the proof. $\square$

DEFINITION 5.1.5. A linear form on $\mathcal{D}_{L^\infty}$ continuous with respect to $\beta$-convergence is said to be an integrable distribution (exactly its restriction to $\mathcal{D}$). The space of all integrable distributions will be denoted by $\mathcal{D}'_{L^1}$.

In accordance with Theorem 2.4.2 and Theorem 2.4.3 every distribution with a compact support is integrable. Let $\Lambda$ be an integrable distribution. We shall now show that the restriction $\Lambda|_{\overset{\circ}{\mathcal{B}}}$ of $\Lambda$ into $\overset{\circ}{\mathcal{B}}$ is continuous under the natural convergence of $\overset{\circ}{\mathcal{B}}$. Indeed, let a sequence $(\varphi_\nu)$, $\varphi_\nu \in \overset{\circ}{\mathcal{B}}$, converge to zero in $\overset{\circ}{\mathcal{B}}$. According to the definition of the convergence in $\overset{\circ}{\mathcal{B}}$ there exists constant $M_\alpha$, $\alpha \in \mathbb{N}^n$ such that $\|\partial^\alpha \varphi_\nu\|_{L^\infty} \leq M_\alpha$, $\nu = 1, 2, \ldots$. Moreover, the sequences $(\partial^\alpha \varphi_\nu)$, $|\alpha| \leq m$ uniformly converge to zero on $\mathbb{R}^n$. This implies that the conditions (i) and (ii) are fulfilled. Since $\mathcal{D} \subset \overset{\circ}{\mathcal{B}} \subset \mathcal{D}_{L^\infty}$ and $\mathcal{D}$ is dense in $\mathcal{D}_{L^\infty}$, we conclude by the Lebesgue dominated convergence theorem that the formula (5.1.1) is true for $\varphi \in \mathcal{D}_{L^\infty}$, too.

Our consideration give us the following

THEOREM 5.1.4. *If $\Lambda \in \mathcal{D}'_{L^1}$, then there exist a nonnegative integer $m$ and complex regular Borel measures $\mu_\alpha$, $|\alpha| \leq m$ such that*

$$\Lambda(\varphi) = \sum_{|\alpha| \leq m} (-1)^{|\alpha|} \int_{\mathbb{R}^n} \partial^\alpha \varphi \, d\mu_\alpha \quad for \quad \varphi \in \mathcal{D}_{L^\infty}. \tag{5.1.2}$$

## 5.2. Regularization of integrable distributions

DEFINITION 5.2.1. We shall denote by $\mathcal{D}_{L^q}$, $1 \leq q \leq \infty$, the set of all smooth functions $\varphi$ such that $\partial^\alpha \varphi \in L^q(\mathbb{R}^n)$ for $\alpha \in \mathbb{N}^n$. Suppose that $\varphi_\epsilon \in \mathcal{D}_{L^q}$, $\epsilon > 0$. We say that the $\varphi_\epsilon$ converge to $\varphi$ in $\mathcal{D}_{L^q}$ if the $\partial^\alpha \varphi_\epsilon$ tend to $\partial^\alpha \varphi$ as $\epsilon \to 0$ in $L^q(\mathbb{R}^n)$ for all $\alpha \in \mathbb{N}^n$.

LEMMA 5.2.1. *If $\varphi \in \mathcal{D}_{L^q}$, $1 \leq q \leq \infty$ and $\psi \in L^1(\mathbb{R}^n)$, then $\varphi * \psi$ is a smooth function and $\partial^\alpha(\varphi * \psi) = \partial^\alpha \varphi * \psi$ for all $\alpha \in \mathbb{N}^n$.*

PROOF. It is known that $\mathcal{D}_{L^{q_1}} \subset \mathcal{D}_{L^{q_2}}$ if $1 \leq q_1 \leq q_2 \leq \infty$ (see Remark 5.4.1). In particular, $\mathcal{D}_{L^q} \subset \mathcal{D}_{L^\infty}$ if $q \geq 1$. Note that

$$h^{-1}((\varphi * \psi)(x + he_i) - (\varphi * \psi)(x)) - (\partial^{e_i} \varphi * \psi)(x)$$
$$= \int_{\mathbb{R}^n} (\partial^{e_i} \varphi(x + \Theta h e_i - y) - \partial^{e_i} \varphi(x - y)) \, \psi(y) \, dy,$$

where $0 < \Theta < 1$. By the Lebesgue dominated convergence theorem we infer that

$$\lim_{h \to 0} \int_{\mathbb{R}^n} (\partial^{e_i} \varphi(x + \Theta h e_i - y) - \partial^{e_i} \varphi(x - y)) \, \psi(y) \, dy = 0$$

for fixed $x \in \mathbb{R}^n$. We now proceed by induction. $\square$

LEMMA 5.2.2. *Let $\psi$ be in $L^1$ and $\int_{\mathbb{R}^n} \psi = 1$. Set $\psi_\epsilon(x) := \epsilon^{-n} \psi\left(\frac{x}{\epsilon}\right)$. If $\varphi \in \mathcal{D}_{L^q}$, $1 \leq q \leq \infty$, then the $\varphi * \psi_\epsilon$ tend to $\varphi$ as $\epsilon \to 0$ in $\mathcal{D}_{L^q}$.*

PROOF. This lemma is an immediate consequence of Lemmas 5.2.1 and 1.2.1. $\square$

THEOREM 5.2.1. *If $\Lambda \in \mathcal{D}'_{L^1}$ and $\varphi \in \mathcal{D}_{L^\infty}$ or $\varphi \in \mathcal{D}_{L^1}$, then*

$$\Lambda_\varphi(x) := \Lambda(\varphi(x - \cdot))$$

*is a smooth function (compare with Theorem 3.1.2). Moreover, $\Lambda_\varphi$ is in $\mathcal{D}_{L^\infty}$ if $\varphi \in \mathcal{D}_{L^\infty}$ and $\Lambda_\varphi$ is in $\mathcal{D}_{L^1}$ if $\varphi \in \mathcal{D}_{L^1}$.*

PROOF. By virtue of Theorem 5.1.4,

$$\Lambda_\varphi(x) = \sum_{|\beta| \leq m} (-1)^{|\beta|} \int_{\mathbb{R}^n} \partial^\beta \varphi(x - \cdot) d\mu_\beta.$$

Note that the $\varphi(x - \cdot)$ tend to $\varphi(x_0 - \cdot)$ as $x \to x_0$ in $\mathcal{D}_{L^\infty}$. This implies that $\Lambda_\varphi$ is continuous in $\mathbb{R}^n$. One can similarly show, as in the proof of Lemma 5.2.1, that

$$\partial^\alpha \Lambda_\varphi(x) = \sum_{|\beta| \leq m} (-1)^{|\beta|} \int_{\mathbb{R}^n} \partial^{\alpha+\beta} \varphi(x - \cdot) d\mu_\beta.$$

Of course, $\partial^\alpha \Lambda_\varphi$ is continuous in $\mathbb{R}^n$. Suppose now that $\varphi$ is in $\mathcal{D}_{L^1}$. In order to prove that $\partial^\alpha \Lambda_\varphi$ is in $\mathcal{D}_{L^1}$ we shall estimate

$$\|\partial^\alpha \Lambda_\varphi\|_{L^1} \le \sum_{|\beta|\le m} \int_{\mathbb{R}^n}\int_{\mathbb{R}^n} |\partial^{\alpha+\beta}\varphi(x-y)|\, dx\, d|\mu_\beta|(y)$$

$$\le M \sum_{|\beta|\le m} |\mu_\beta|(\mathbb{R}^n),$$

where $M = \max_{|\beta|\le m} \|\partial^{\alpha+\beta}\varphi\|_{L^1}$. Analogously one can show that

$$\|\partial^\alpha \Lambda_\varphi\|_{L^\infty} \le M_1 \sum_{|\beta|\le m} |\mu_\beta|(\mathbb{R}^n),$$

where $M_1 = \max_{|\beta|\le m} \|\partial^{\alpha+\beta}\varphi\|_{L^\infty}$.    $\square$

DEFINITION 5.2.2. The function $\Lambda_\varphi$ is called the regularization of $\Lambda$ by means of $\varphi$.

## 5.3. Tensor product of integrable distributions

Let $\varphi$ be in $\mathcal{D}_{L^\infty}(\mathbb{R}^m \times \mathbb{R}^n)$. Assume that $S \in \mathcal{D}'_{L^1}(\mathbb{R}^m)$ and $T \in \mathcal{D}'_{L^1}(\mathbb{R}^n)$. In much the same way as in the proof of Theorem 5.2.1 one can show that

$$\psi(x) := T(\varphi(x,\cdot))$$

belongs to $\mathcal{D}_{L^\infty}(\mathbb{R}^m)$. Therefore the expression $S_x(T_y(\varphi(x,y))$ is sensible for $\varphi \in \mathcal{D}_{L^\infty}(\mathbb{R}^m \times \mathbb{R}^n)$. Moreover, according to Theorem 5.1.4 we have

$$S_x(T_y(\varphi(x,y)))$$
$$= \sum_{|\alpha|\le m_1} (-1)^{|\alpha|} \int_{\mathbb{R}^m} \frac{\partial^{|\alpha|}}{\partial x^\alpha}\left( \sum_{|\beta|\le m_2} (-1)^{|\beta|} \int_{\mathbb{R}^n} \frac{\partial^{|\beta|}}{\partial y^\beta}\varphi(x,y)\, d\mu_\beta(y) \right) d\mu_\alpha(x)$$
$$= \sum_{|\alpha|\le m_1}\sum_{|\beta|\le m_2} (-1)^{|\alpha+\beta|} \int_{\mathbb{R}^m}\left( \int_{\mathbb{R}^n} \frac{\partial^{|\alpha+\beta|}}{\partial x^\alpha \partial y^\beta}\varphi(x,y)h_\alpha(x)h_\beta(y)\, d|\mu_\beta|(y) \right) d|\mu_\alpha|(x)$$

$$(5.3.1)$$

(see Section 2.7). Note that the mapping

$$\mathbb{R}^m \times \mathbb{R}^n \ni (x,y) \to \frac{\partial^{|\alpha+\beta|}}{\partial x^\alpha \partial y^\beta}\varphi(x,y)h_\alpha(x)h_\beta(y)$$

is a bounded Borel function on $\mathbb{R}^m \times \mathbb{R}^n$. Therefore, by the Fubini theorem ([23]) we obtain the following equality

$$\int_{\mathbb{R}^m}\left( \int_{\mathbb{R}^n} \frac{\partial^{|\alpha+\beta|}}{\partial x^\alpha \partial y^\beta}\varphi(x,y)h_\alpha(x)h_\beta(y)\, d|\mu_\beta|(y) \right) d|\mu_\alpha|(x)$$
$$= \int_{\mathbb{R}^m \times \mathbb{R}^n} \frac{\partial^{|\alpha+\beta|}}{\partial x^\alpha \partial y^\beta}\varphi(x,y)h_\alpha(x)h_\beta(y)\, d(|\mu_\alpha|\times|\mu_\beta|)(x,y),$$

where $|\mu_\alpha| \times |\mu_\beta|$ denotes the product of the measures $|\mu_\alpha|$ and $|\mu_\beta|$. It is convenient to write the double integral $\int_{\mathbb{R}^m} \int_{\mathbb{R}^n} \frac{\partial^{|\alpha+\beta|}}{\partial x^\alpha \partial y^\beta} \varphi(x,y)\, d\mu_\alpha(x)\, d\mu_\beta(y)$ in the place of $\int_{\mathbb{R}^m \times \mathbb{R}^n} \frac{\partial^{|\alpha+\beta|}}{\partial x^\alpha \partial y^\beta} \varphi(x,y) h_\alpha(x) h_\beta(y)\, d(|\mu_\alpha| \times |\mu_\beta|)(x,y)$. Thus, we have

$$S_x(T_y(\varphi(x,y))) = \sum_{|\alpha|\leq m_1} \sum_{|\beta|\leq m_2} (-1)^{|\alpha+\beta|} \int_{\mathbb{R}^m} \int_{\mathbb{R}^n} \frac{\partial^{|\alpha+\beta|}}{\partial x^\alpha \partial y^\beta} \varphi(x,y)\, d\mu_\alpha(x)\, d\mu_\beta(y).$$

Note that the mapping

$$\mathcal{D}_{L^\infty}(\mathbb{R}^m \times \mathbb{R}^n) \ni \varphi \longrightarrow S_x(T_y(\varphi(x,y)))$$

is linear and continuous with respect to $\beta$-convergence (see Definition 5.1.4). In accordance with the general definition of tensor product of two arbitrary distributions put

$$(S \otimes T)(\varphi) := S_x(T_y(\varphi(x,y))).$$

It follows from the above consideration that $S \otimes T$ is in $\mathcal{D}'_{L^1}(\mathbb{R}^m \times \mathbb{R}^n)$. Obviously, $S \otimes T = T \otimes S$.

DEFINITION 5.3.1. The distribution $S \otimes T$ given by (5.3.1) is said to be the tensor product of $S$ and $T$.

Of course, if $S$ and $T$ belong to $\mathcal{D}'_{L^1}(\mathbb{R}^n)$, then the mapping

$$\mathcal{D}_{L^\infty}(\mathbb{R}^n) \ni \varphi \longrightarrow (S_x \otimes T_y)\varphi(x+y)$$

is an integrable distribution on $\mathbb{R}^n$. We shall denote this distribution by $S * T$ and call the convolution product of $S$ and $T$. Taking into account (5.3.1) one can verify that

$$(S * T) * V = S * (T * V)$$

for $S$, $T$ and $V$ belonging to $\mathcal{D}'_{L^1}(\mathbb{R}^n)$. We are able to prove the following

THEOREM 5.3.1. If $\Lambda \in \mathcal{D}'_{L^1}$ and $f \in \mathcal{D}_{L^1}$, then

$$(\Lambda * f)(\varphi) = \int_{\mathbb{R}^n} \Lambda_f(x)\varphi(x)\, dx$$

for $\varphi \in \mathcal{D}$.

PROOF. We have to show that

$$(\Lambda_y \otimes f_x)\varphi(x+y) = \int_{\mathbb{R}^n} \Lambda_f(x)\varphi(x)\, dx$$

for $\varphi \in \mathcal{D}$. The following equalities

$$(\Lambda_y \otimes f_x)\varphi(x+y) = \Lambda_y \left( \int_{\mathbb{R}^n} f(x)\varphi(x+y)\, dx \right) = \Lambda_y \left( \int_{\mathbb{R}^n} f(\omega - y)\varphi(\omega)\, d\omega \right)$$

are evident. As in the proof of Theorem 3.1.4 we put

$$\sigma_h(y) := \sum_{\nu \in Z_h^n} f(h\nu - y)\varphi(h\nu)h^n,$$

where $Z_h^n = \{\nu : h\nu \in \operatorname{supp} \varphi\}$. One can show that $\sigma_h(y)$ tends to

$$\int\limits_{\mathbb{R}^n} f(\omega - y)\varphi(\omega)\,d\omega \quad \text{as } h \to 0$$

in $\mathcal{D}_{L^\infty}$. This implies that the $\Lambda(\sigma_h(\cdot))$ tend to

$$\Lambda_y \left( \int\limits_{\mathbb{R}^n} f(\omega - y)\varphi(\omega)\,d\omega \right) \quad \text{as } h \to 0.$$

On the other hand we have

$$\Lambda(\sigma_h(\cdot)) = \sum_{\nu \in Z_h^n} \Lambda(f(h\nu - \cdot)\varphi(h\nu))h^n.$$

The right side of this equality is a Riemann sum of the integral $\int_{\mathbb{R}^n} \Lambda(f(\omega - \cdot)\varphi(\omega))\,d\omega$. Finally we obtain

$$(\Lambda_y \otimes f_x)\varphi(x + y) = \int\limits_{\mathbb{R}^n} \Lambda_f(\omega)\varphi(\omega)\,d\omega.$$

Thus, the theorem is proved. $\qquad\square$

The next theorem is very important for applications.

THEOREM 5.3.2. *If $\Lambda \in \mathcal{D}'_{L^1}$, $\psi \in \mathcal{D}_{L^1}$ and $\int_{\mathbb{R}^n} \psi = 1$, then the $\int_{\mathbb{R}^n} \Lambda_{\psi_\epsilon}(x)\varphi(x)\,dx$ tend to $\Lambda(\varphi)$ as $\epsilon \to 0$ for $\varphi \in \mathcal{D}_{L^\infty}$, where $\psi_\epsilon(x) = \epsilon^{-n}\psi\left(\frac{x}{\epsilon}\right)$.*

PROOF. By virtue of Theorem 5.2.1, $\Lambda_{\psi_\epsilon} \in L^1(\mathbb{R}^n)$. Hence for $\varphi \in \mathcal{D}_{L^\infty}$ we have

$$\int\limits_{\mathbb{R}^n} \Lambda_{\psi_\epsilon}(x)\varphi(x)\,dx = \int\limits_{\mathbb{R}^n} \left( \sum_{|\beta| \leq m} \int\limits_{\mathbb{R}^n} \partial^\beta \psi_\epsilon(x - y)\,d\mu_\beta(y) \right) \varphi(x)\,dx$$

$$= \sum_{|\beta| \leq m} \int\limits_{\mathbb{R}^n} \left( \int\limits_{\mathbb{R}^n} \partial^\beta \psi_\epsilon(x - y)\varphi(x)\,dx \right) d\mu_\beta(y)$$

$$= \sum_{|\beta| \leq m} (-1)^{|\beta|} \int\limits_{\mathbb{R}^n} \left( \int\limits_{\mathbb{R}^n} \psi_\epsilon(x - y)\partial^\beta\varphi(x)\,dx \right) d\mu_\beta(y).$$

After a change of variable in the interior integral we get

$$\int\limits_{\mathbb{R}^n} \Lambda_{\psi_\epsilon}(x)\varphi(x)\,dx = \sum_{|\beta| \leq m} (-1)^{|\beta|} \int\limits_{\mathbb{R}^n} \left( \int\limits_{\mathbb{R}^n} \psi(\omega)\partial^\beta\varphi(y + \epsilon\omega)\,d\omega \right) d\mu_\beta(y).$$

On the other hand we have

$$\Lambda(\varphi) = \sum_{|\beta| \leq m} (-1)^{|\beta|} \int\limits_{\mathbb{R}^n} \left( \int\limits_{\mathbb{R}^n} \psi(\omega)\,d\omega \right) \partial^\beta\varphi(y)\,d\mu_\beta(y).$$

Therefore

$$\int\limits_{\mathbb{R}^n} \Lambda_{\psi_\epsilon}(x)\varphi(x)\,dx - \Lambda(\varphi)$$

$$= \sum_{|\beta|\leq m} (-1)^{|\beta|} \int\limits_{\mathbb{R}^n}\int\limits_{\mathbb{R}^n} \left( \partial^\beta\varphi(y+\epsilon\omega) - \partial^\beta\varphi(y) \right)\psi(\omega)\,d\mu_\beta(y)\,d\omega.$$

Note that for fixed $\omega$ the expression $\partial^\beta\varphi(y+\epsilon\omega) - \partial^\beta\varphi(y)$ uniformly converges to zero as $\epsilon \to 0$ with respect to $y$. Of course, this expression is bounded on $\mathbb{R}^n \times \mathbb{R}^n$. Because $\psi \in L^1(\mathbb{R}^n)$, we can use the Lebesgue dominated convergence theorem. Therefore the

$$\int\limits_{\mathbb{R}^n}\int\limits_{\mathbb{R}^n} \left( \partial^\beta\varphi(y+\epsilon\omega) - \partial^\beta\varphi(y) \right)\psi(\omega)\,d\mu_\beta(y)\,d\omega$$

tend to zero as $\epsilon \to 0$. This statement finishes the proof.               $\square$

### 5.4. $\mathcal{D}_{L^p}$ and $\mathcal{D}'_{L^p}$ spaces

In this section we shall examine a new class of distributions containing the set of integrable distributions. As the test space for these distributions we take the space $\mathcal{D}_{L^q}$, $1 < q < \infty$ given by Definition 5.2.1. The convergence in $\mathcal{D}_{L^q}$ may be defined by means of the following family of norms

$$\|\varphi\|_m = \left( \sum_{|\alpha|\leq m} \|\partial^\alpha\varphi\|_{L^q}^q \right)^{\frac{1}{q}}, \qquad m = 0, 1, 2, \ldots. \qquad (5.4.1)$$

DEFINITION 5.4.1. Let $1 < q < \infty$ and $\frac{1}{p} + \frac{1}{q} = 1$. The set of all linear forms defined on $\mathcal{D}_{L^q}$ and continuous with respect to the norms (5.4.1) will be denoted by $\mathcal{D}'_{L^p}$.

We shall now appeal to the well-known facts concerning Sobolev spaces, which will be collected in the following

REMARK 5.4.1. In view of the Sobolev imbedding theorem ([1]), it follows that $W^{m+1,q_1} \subset W^{m,q_2}$, $1 \leq q_1 \leq q_2 < \infty$ and $\|\varphi\|_{m,q_2} \leq \|\varphi\|_{m+1,q_1}$ for $\varphi \in W^{m+1,q_1}$. This means that the natural imbedding of $\mathcal{D}_{L^{q_1}}$ into $\mathcal{D}_{L^{q_2}}$ is continuous if $1 \leq q_1 \leq q_2 < \infty$. As a consequence of this fact, we have $\mathcal{D}'_{L^{p_1}} \subset \mathcal{D}'_{L^{p_2}}$ for $1 < p_1 \leq p_2 \leq \infty$ (this means that the restriction $\Lambda|_{\mathcal{D}_{L^{q_1}}}$ of $\Lambda \in \mathcal{D}'_{L^{p_2}}$ into $\mathcal{D}_{L^{q_1}}$, $\frac{1}{p_1} + \frac{1}{q_1} = 1$ belongs to $\mathcal{D}'_{L^{p_1}}$). It will be also shown that $\mathcal{D}'_{L^1} \subset \mathcal{D}'_{L^p}$ for $p \geq 1$ (see Remark 5.6.1).

In accordance with Lemma 1.12.1, $\mathcal{D}$ is dense in $\mathcal{D}_{L^q}$, $1 \leq q < \infty$. Therefore, a distribution $\Lambda$ which is continuous with respect to some norm of family (5.4.1) may be extended by continuity on $\mathcal{D}_{L^q}$. We are able to formulate a representation theorem for the elements belonging to $\mathcal{D}'_{L^p}$.

THEOREM 5.4.1. *Each element $\Lambda$ belonging to $\mathcal{D}'_{L^p}$, $1 < p < \infty$ takes the following form*

$$\Lambda(\varphi) = \sum_{|\alpha|\leq m} (-1)^{|\alpha|} \int\limits_{\mathbb{R}^n} f_\alpha \partial^\alpha\varphi,$$

*where $\varphi \in \mathcal{D}_{L^q}$, $\frac{1}{p} + \frac{1}{q} = 1$ and $f_\alpha$ is a fixed element of $L^p$.*

PROOF. This theorem may be proved in much the same way as Theorem 2.7.3. In this case we only need to take the Cartesian product $\mathcal{L}^N(\mathbb{R}^n) := X_{|\alpha| \le m} L^q(\mathbb{R}^n)$, where $N = \sum_{|\alpha| \le m} 1$ instead of $C^N(\Omega') = X_{|\alpha| \le m} C_0(\Omega')$ and to replace Theorem 2.7.2 by the Riesz representation theorem for bounded linear forms on $L^q(\mathbb{R}^n)$.    □

## 5.5. Convolution product

We begin our consideration with presenting an important theorem concerning convolution product of functions.

THEOREM 5.5.1 (Young, [21, vol.2]. ] If $f \in L^p(\mathbb{R}^n)$ and $g \in L^q(\mathbb{R}^n)$, $1 \le p$, $q \le \infty$ and $\frac{1}{p} + \frac{1}{q} - 1 \ge 0$, then the convolution product $f * g$ is in $L^r(\mathbb{R}^n)$, where $\frac{1}{p} + \frac{1}{q} - 1 = \frac{1}{r}$ and

$$\|f * g\|_{L^r} \le \|f\|_{L^p} \|g\|_{L^q}. \tag{5.5.1}$$

We shall now show that a similar theorem is true if the distributions $S$ and $T$ are in $\mathcal{D}'_{L^p}$ and $\mathcal{D}'_{L^q}$, respectively. It was shown in Section 5.3, that if $S$ and $T$ belong to $\mathcal{D}'_{L^1}$, then their convolution product is also in $\mathcal{D}'_{L^1}$. Now, suppose that $S$ is in $\mathcal{D}'_{L^1}$ and $T$ is in $\mathcal{D}'_{L^p}$, $1 < p < \infty$. We define similarly to Section 5.3 the convolution product $S * T$ of $S$ and $T$ as follows

$$(S * T)(\varphi) = (S_x \otimes T_y)\varphi(x + y). \tag{5.5.2}$$

We shall now show that the above formula is meaningful for $\varphi \in \mathcal{D}_{L^q}$, $\frac{1}{p} + \frac{1}{q} = 1$. Moreover, it will be shown that the mapping

$$\mathcal{D}_{L^q} \ni \varphi \longrightarrow (S * T)(\varphi) \tag{5.5.3}$$

is continuous. Formally, by virtue of Theorems 5.1.4 and 5.4.1 we can write

$$(S_x \otimes T_y)\varphi(x + y)$$
$$= \sum_{|\alpha| \le m_1} \sum_{|\beta| \le m_2} (-1)^{|\alpha + \beta|} \int_{\mathbb{R}^n} \int_{\mathbb{R}^n} h_\alpha(x) f_\beta(y) \partial^{\alpha + \beta} \varphi(x + y) \, dy \, d|\mu_\alpha|(x),$$

where $f_\beta \in L^p(\mathbb{R}^n)$, $|h_\alpha(x)| = 1$ for $x \in \mathbb{R}^n$ and $|\mu_\alpha|$ is the total variation of $\mu_\alpha$ (see Section 2.7). Now, we have to prove that there exists the double integral

$$\int_{\mathbb{R}^n} \int_{\mathbb{R}^n} h_\alpha(x) f_\beta(y) \partial^{\alpha + \beta} \varphi(x + y) \, dy \, d|\mu_\alpha|(x). \tag{5.5.4}$$

Put

$$\psi_\beta(x) := \int_{\mathbb{R}^n} f_\beta(y) \partial^{\alpha + \beta} \varphi(x + y) \, dy.$$

Since $h_\alpha$ is bounded in $\mathbb{R}^n$, it suffices to show that $\psi_\beta$ is also bounded in $\mathbb{R}^n$. It is easy to verify that $\psi_\beta(x) = \check{f}_\beta * \partial^{\alpha + \beta} \varphi$, where $\check{f}_\beta(\cdot) = f_\beta(-\cdot)$. By Theorem 5.5.1 we have

$$\|\psi_\beta\|_{L^\infty} \le \|f_\beta\|_{L^p} \|\partial^{\alpha + \beta} \varphi\|_{L^q}.$$

Therefore there exists the integral $\int_{\mathbb{R}^n} |h_\alpha(x)\psi_\beta(x)|\, d|\mu_\alpha|(x)$. This implies the existence of (5.5.4). From this, by Fubini's theorem we infer that the equality

$$(S * T)(\varphi) = (S_x \otimes T_y)\varphi(x + y)$$

is sensible for $\varphi \in L^q$ and

$$|(S * T)(\varphi)| \leq \sum_{|\alpha| \leq m_1} \sum_{|\beta| \leq m_2} \|f_\beta\|_{L^p} |\mu_\alpha|(\mathbb{R}^n) \|\partial^{\alpha+\beta}\varphi\|_{L^q}.$$

This implies that (5.5.3) is continuous.

It remains to consider the third case, when $1 < p, q$ and $\frac{1}{p} + \frac{1}{q} - 1 \geq 0$. In this case we can write formally, by virtue of Theorem 5.4.1

$$(S * T)(\varphi)$$

$$= \sum_{|\alpha| \leq m_1} \sum_{|\beta| \leq m_2} (-1)^{|\alpha+\beta|} \int_{\mathbb{R}^n} \int_{\mathbb{R}^n} f_\alpha(x) g_\beta(y) \partial^{\alpha+\beta}\varphi(x+y)\, dy\, dx, \quad (5.5.5)$$

where $f_\alpha \in L^p(\mathbb{R}^n)$ and $g_\beta \in L^q(\mathbb{R}^n)$. We shall now prove that the right side of (5.5.5) is sensible for $\varphi \in \mathcal{D}_{L^s}$, where $\frac{1}{p} + \frac{1}{q} + \frac{1}{s} = 2$. Further, it will be shown that the mapping

$$\mathcal{D}_{L^s} \ni \varphi \longrightarrow (S * T)(\varphi) \qquad (5.5.6)$$

is continuous. Similarly as in the above case we set

$$\psi_\beta(x) := \int_{\mathbb{R}^n} g_\beta(y) \partial^{\alpha+\beta}\varphi(x+y)\, dy.$$

Since $g_\beta$ is in $L^q(\mathbb{R}^n)$ and $\partial^{\alpha+\beta}\varphi \in L^s(\mathbb{R}^n)$, it follows by Theorem 5.5.1 that $\psi_\beta \in L^t(\mathbb{R}^n)$, where $\frac{1}{s} + \frac{1}{q} - 1 = \frac{1}{t}$. It is easily seen that $\frac{1}{t} = 1 - \frac{1}{p}$. Hence we have $\frac{1}{t} + \frac{1}{p} = 1$. In view of Hölder's inequality we obtain

$$\int_{\mathbb{R}^n} |f_\alpha(x)\psi_\beta(x)|\, dx \leq \|f_\alpha\|_{L^p} \|\psi_\beta\|_{L^t}.$$

This implies the existence of the double integral

$$\int_{\mathbb{R}^n} \int_{\mathbb{R}^n} f_\alpha(x) g_\beta(y) \partial^{\alpha+\beta}\varphi(x+y)\, dx\, dy.$$

It is easy to see that

$$|(S * T)(\varphi)| \leq \sum_{|\alpha| \leq m_1} \sum_{|\beta| \leq m_2} \|f_\alpha\|_{L^p} \|g_\beta\|_{L^q} \|\partial^{\alpha+\beta}\varphi\|_{L^s}.$$

This means that (5.5.6) is continuous. Therefore $S * T$ is in $\mathcal{D}'_{L^r}$, where $\frac{1}{r} + \frac{1}{s} = 1$. It is easy to check that $\frac{1}{p} + \frac{1}{q} - 1 = \frac{1}{r}$.

Finally we obtain the following

THEOREM 5.5.2 (L. Schwartz). *If $S$ and $T$ belong to $\mathcal{D}'_{L^p}$ and $\mathcal{D}'_{L^q}$ respectively, $1 \leq p, q < \infty$ and $\frac{1}{p} + \frac{1}{q} - 1 \geq 0$, then there exists the convolution product $S * T$ and $S * T \in \mathcal{D}'_{L^r}$, where $\frac{1}{p} + \frac{1}{q} - 1 = \frac{1}{r}$.*

## 5.6. Cauchy transforms of integrable distributions

It is easily seen that the function $\frac{1}{2\pi i}\frac{1}{(\cdot - z)}$, $\Im z \neq 0$ is in $\mathcal{D}_{L^\infty}(\mathbb{R})$. This remark makes the following symbol

$$C\Lambda(z) := \frac{1}{2\pi i}\Lambda\left(\frac{1}{\cdot - z}\right) \qquad (5.6.1)$$

sensible for $\Lambda \in \mathcal{D}'_{L^1}$ and $\Im z \neq 0$.

DEFINITION 5.6.1. The function $C\Lambda$ is said to be the Cauchy transform of $\Lambda$.

We shall now prove the following

THEOREM 5.6.1. If $\Lambda \in \mathcal{D}'_{L^1}(\mathbb{R})$, then $C\Lambda$ is a holomorphic function in $\mathbb{C} - \mathbb{R}$ and

$$\frac{d}{dz}C\Lambda(z) = \frac{1}{2\pi i}\Lambda_t\left(\frac{1}{(t-z)^2}\right) = C\Lambda^{(1)}(z), \qquad (5.6.2)$$

where $\Lambda^{(1)}$ is the distributional derivative of $\Lambda$.

PROOF. To this end we shall show that

$$\lim_{h \to 0} \frac{1}{2\pi i}\Lambda_t\left(h^{-1}((t-z-h)^{-1} - (t-z)^{-1}) - (t-z)^{-2}\right) = 0.$$

To prove this one we only need to verify that for fixed $z$ and $\alpha \in \mathbb{N}$, the $\frac{\partial^\alpha}{\partial t^\alpha}(h(t-z)^{-2}(t-z-h)^{-1})$ tend uniformly to zero on $\mathbb{R}$ as $h \to 0$. For simplicity of notation put

$$g_{h,z}(t) := h(t-z)^{-2}(t-z-h)^{-1}.$$

By the Leibniz formula we get

$$\frac{\partial^\alpha}{\partial t^\alpha}g_{h,z}(t) = (-1)^\alpha \alpha! h \sum_{\nu=0}^{\alpha}(\nu+1)(t-z)^{-(\nu+2)}(t-z-h)^{-(\alpha+1-\nu)}.$$

Putting $z = x + iy$, $h = \xi + i\eta$ we obtain

$$\left|\frac{\partial^\alpha}{\partial t^\alpha}g_{h,z}(t)\right|$$
$$\leq \alpha!|h|\sum_{\nu=0}^{\alpha}(\nu+1)((t-x)^2+y^2)^{-\frac{\nu+2}{2}}((t-x-\xi)^2+(y+\eta)^2)^{-\frac{\alpha+1-\nu}{2}}. \quad (5.6.3)$$

Set $d = y - \delta$ if $y > 0$ and $d = |y + \delta|$ if $y < 0$. Hence, for $|h| < \delta < |y|$ we have

$$((t-x)^2+y^2)^{-\frac{\nu+2}{2}}((t-x-\xi)^2+(y+\eta)^2)^{-\frac{\alpha+1-\nu}{2}} \leq d^{-\alpha-3}$$

for $t \in \mathbb{R}$. Finally we obtain the following estimation

$$\left|\frac{\partial^\alpha}{\partial t^\alpha}g_{h,z}(t)\right| \leq C|h|d^{-\alpha-3}, \qquad C = \alpha!\sum_{\nu=0}^{\alpha}(\nu+1).$$

Thus, we have proved that the $\frac{\partial^\alpha}{\partial t^\alpha}g_{h,z}(t)$ converge uniformly to zero with respect to $t \in \mathbb{R}$ as $h \to 0$. Hence, it follows that

$$\frac{d}{dz}C\Lambda(z) = \frac{1}{2\pi i}\Lambda\left(\frac{1}{(\cdot - z)^2}\right).$$

Note that

$$\frac{1}{2\pi i}\Lambda^{(1)}\left(\frac{1}{\cdot - z}\right) = \frac{-1}{2\pi i}\Lambda_t\left(\frac{\partial}{\partial t}\frac{1}{t-z}\right) = \frac{1}{2\pi i}\Lambda\left(\frac{1}{(\cdot - z)^2}\right).$$

Finally, we obtain (5.6.2). □

The next theorem is of theoretical interest and is important for applications.

THEOREM 5.6.2. *If $f \in L^1(\mathbb{R})$, then*

$$Cf(\cdot + i\epsilon) - Cf(\cdot - i\epsilon) \to f \text{ as } \epsilon \to 0 \text{ in } L^1(\mathbb{R}). \tag{5.6.4}$$

PROOF. An easy computation shows that

$$Cf(\cdot + i\epsilon) - Cf(\cdot - i\epsilon) = h_\epsilon * f, \tag{5.6.5}$$

where $h_\epsilon = \frac{1}{\epsilon}h\left(\frac{t}{\epsilon}\right)$, $h(t) = \frac{1}{\pi}\frac{1}{1+t^2}$. By Lemma 1.2.1 we infer that $Cf(\cdot+i\epsilon)-Cf(\cdot - i\epsilon)$ tends to $f$ as $\epsilon \to 0$ in $L^1(\mathbb{R})$. This finishes the proof. □

COROLLARY 5.6.1. *If $f$ is in $\mathcal{D}_{L^1}$ then*

$$Cf(\cdot + i\epsilon) - Cf(\cdot - i\epsilon) \to f \text{ as } \epsilon \to 0 \text{ in } \mathcal{D}_{L^1}(\mathbb{R}). \tag{5.6.6}$$

This corollary is an immediate consequence of Lemma 5.2.2.

EXAMPLE 5.6.1. Put $f(t) := \frac{1}{t^2+1}$. We shall apply the theory of residues to determining $Cf$. Let us consider the following contour

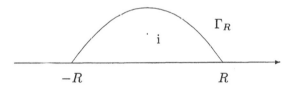

It is easy to check that

$$\lim_{R\to\infty}\frac{1}{2\pi i}\int_{\Gamma_R}\frac{1}{(t^2+1)(t-z)}\,dt = 0.$$

Note that

$$Cf(z) = \lim_{R\to\infty}\frac{1}{2\pi i}\int_{[-R,R]\cup\Gamma_R}\frac{1}{(t^2+1)(t-z)}\,dt = \frac{1}{2i(i-z)}$$

if $\Im z < 0$. Similarly one can show that

$$Cf(z) = \frac{-1}{2i(i+z)}$$

if $\Im z > 0$. According to Theorem 5.6.2

$$\frac{-1}{2i(\cdot + i(1+\epsilon))} + \frac{1}{2i(\cdot - i(1+\epsilon))} \to \frac{1}{(\cdot)^2+1}$$

in $L^1(\mathbb{R})$.

A similar theorem to Theorem 5.6.2 is true if $\Lambda$ is an integrable distribution.

THEOREM 5.6.3. *If $\Lambda$ belongs to $\mathcal{D}'_{L^1}$, then the $C\Lambda(\cdot + i\epsilon) - C\Lambda(\cdot - i\epsilon)$ tend to $\Lambda$ as $\epsilon \to 0$ in $\mathcal{D}'_{L^1}$.*

PROOF. In accordance with Theorem 5.1.4 the integrable distribution $\Lambda$ can be written as follows

$$\Lambda(\varphi) = \sum_{\alpha=0}^{m} (-1)^\alpha \int_{\mathbb{R}} \partial^\alpha \varphi \, d\mu_\alpha,$$

where $\mu_\alpha$ is a complex regular Borel measure and $\varphi \in \mathcal{D}_{L^\infty}$. We have to show that

$$\int_{\mathbb{R}} (C\Lambda(x + i\epsilon) - C\Lambda(x - i\epsilon))\varphi(x) \, dx \to \sum_{\alpha=0}^{m} (-1)^\alpha \int_{\mathbb{R}} \partial^\alpha \varphi \, d\mu_\alpha$$

as $\epsilon \to 0$. Note that

$$\int_{\mathbb{R}} (C\Lambda(x+i\epsilon) - C\Lambda(x-i\epsilon))\varphi(x) \, dx = \sum_{\alpha=0}^{m} (-1)^\alpha \int_{\mathbb{R}} \left( \int_{\mathbb{R}} \partial^\alpha h_\epsilon(t - x) d\mu_\alpha(t) \right) \varphi(x) \, dx,$$

where $h_\epsilon(t) = \frac{1}{\pi} \frac{\epsilon}{t^2 + \epsilon^2}$. By the Fubini theorem we have

$$\int_{\mathbb{R}} (C\Lambda(x + i\epsilon) - C\Lambda(x - i\epsilon))\varphi(x) \, dx = \sum_{\alpha=0}^{m} (-1)^\alpha \int_{\mathbb{R}} (\partial^\alpha h_\epsilon * \varphi)(t) \, d\mu_\alpha(t)$$

$$= \sum_{\alpha=0}^{m} (-1)^\alpha \int_{\mathbb{R}} (h_\epsilon * \partial^\alpha \varphi)(t) \, d\mu_\alpha(t).$$

Hence, in view of Lemma 5.2.2

$$\lim_{\epsilon \to 0} \sum_{\alpha=0}^{m} (-1)^\alpha \int_{\mathbb{R}} (h_\epsilon * \partial^\alpha \varphi)(t) \, d\mu_\alpha(t) = \sum_{\alpha=0}^{m} (-1)^\alpha \int_{\mathbb{R}} \partial^\alpha \varphi \, d\mu_\alpha.$$

This statement finishes the proof.    $\square$

REMARK 5.6.1. Let $\Lambda$ be in $\mathcal{D}'_{L^1}$. We shall show now that the restriction $\Lambda|_{\mathcal{D}_{L^q}}$ of $\Lambda$ to $\mathcal{D}_{L^q}$, $q > 1$ belongs to $\mathcal{D}'_{L^p}$, where $\frac{1}{p} + \frac{1}{q} = 1$. Indeed, because $\mathcal{D}_{L^q} \subset \mathcal{D}_{L^\infty}$ and $\mathcal{D}_{L^q}$ is complete, therefore by Theorem A.1 (see Appendix) and Theorem 5.6.3 we infer that $\Lambda|_{\mathcal{D}_{L^q}}$ belongs to $\mathcal{D}'_{L^p}$.

The following theorem gives a connection between the Cauchy transforms of integrable distributions and harmonic functions defined in the upper half plane.

THEOREM 5.6.4. *Let $\Lambda$ be in $\mathcal{D}'_{L^1}$, then the function*

$$(x, y) \longrightarrow C\Lambda(x + iy) - C\Lambda(x - iy)$$

*is a harmonic function in $\Pi^+ = \{(x, y) : x \in \mathbb{R}, \ y > 0\}$ and*

$$\lim_{y \to 0^+} (C\Lambda(x + iy) - C\Lambda(x - iy)) = \Lambda \text{ in } \mathcal{D}'_{L^1}.$$

PROOF. Since $\mathbb{C} - \mathbb{R} \ni z \longrightarrow C\Lambda(z)$ is a holomorphic function in $\mathbb{C} - \mathbb{R}$, therefore the function

$$\Pi^+ \ni (x, y) \longrightarrow C\Lambda(x + iy) - C\Lambda(x - iy)$$

is a harmonic function. Theorem 5.6.3 gives us the second part of our theorem. $\square$

This theorem tells that every integrable distribution $\Lambda$ is the distributional boundary value of a harmonic function. The problem of finding a harmonic function, when given boundary values are assumed is known as the Dirichlet problem.

REMARK 5.6.2. It is easily seen that

$$C\Lambda(x + iy) - C\Lambda(x - iy) = \Lambda_t \left( \frac{1}{\pi} \frac{y}{(t - x)^2 + y^2} \right).$$

The formula

$$u(x, y) = \Lambda_t \left( \frac{1}{\pi} \frac{y}{(t - x)^2 + y^2} \right), \tag{5.6.7}$$

where $y > 0$ is a natural generalization of the Poisson integral

$$u(x, y) = \frac{1}{\pi} \int_{\mathbb{R}} \frac{y}{(t - x)^2 + y^2} f(t) dt$$

for the upper half plane.

EXAMPLE 5.6.2. It is easily seen that $C\delta(z) = -\frac{1}{2\pi i z}$, $z \neq 0$, where $\delta$ denotes the Dirac measure concentrated at zero. In accordance with Theorem 5.6.4 the function

$$(x, y) \longrightarrow C\delta(x + iy) - C\delta(x - iy) = \frac{1}{\pi} \frac{y}{x^2 + y^2}$$

is harmonic in $\Pi^+$ and $\frac{1}{\pi} \frac{y}{(\cdot)^2 + y^2}$ tends to $\delta$ as $y \to 0^+$ in $\mathcal{D}'_{L^1}$.

## 5.7. Cauchy transforms of distributions belonging to $\mathcal{D}'_{L^p}$

We shall now show that $\frac{1}{(\cdot - z)}$ is in $\mathcal{D}_{L^q}$, $1 < q < \infty$ if $\Im z \neq 0$. Note that

$$\frac{\partial^\alpha}{\partial t^\alpha}(t - z)^{-1} = (-1)^\alpha \alpha! (t - z)^{-\alpha - 1}.$$

A simple computation shows that

$$\int_{\mathbb{R}} |t - x - iy|^{-q(\alpha+1)} \, dt \leq |y|^{-q(\alpha+1)} \int_{\mathbb{R}} \left( \left( \frac{t - x}{y} \right)^2 + 1 \right)^{-\frac{q}{2}(\alpha+1)} dt$$

$$= |y|^{-q(\alpha+1)+1} \int_{\mathbb{R}} (\omega^2 + 1)^{-\frac{q}{2}(\alpha+1)} \, d\omega$$

$$= C|y|^{-q(\alpha+1)+1},$$

where $C = \int_{\mathbb{R}} (\omega^2 + 1)^{-\frac{q}{2}(\alpha+1)} \, d\omega$. Therefore $\frac{1}{(\cdot - z)} \in \mathcal{D}_{L^q}$, $1 < q < \infty$ if $\Im z \neq 0$. This implies that the symbol $C\Lambda(z) = \frac{1}{2\pi i} \Lambda \left( \frac{1}{\cdot - z} \right)$ is also sensible if $\Lambda \in \mathcal{D}'_{L^p}$, $1 < p < \infty$.

THEOREM 5.7.1. *If $\Lambda \in \mathcal{D}'_{L^p}$, $1 < p < \infty$, then the Cauchy transform $C\Lambda$ is a holomorphic function in $\mathbb{C} - \mathbb{R}$ and*

$$\frac{d}{dz}C\Lambda(z) = \frac{1}{2\pi i}\Lambda\left(\frac{1}{(\cdot - z)^2}\right) = C\Lambda^{(1)}(z),$$

*where $\Lambda^{(1)}$ is the distributional derivative of $\Lambda$.*

PROOF. Analysis similar to that in the proof of Theorem 5.6.1 shows that

$$(C\Lambda(z+h) - C\Lambda(z))h^{-1} - \frac{1}{2\pi i}\Lambda\left(\frac{1}{(\cdot - z)^2}\right) = \frac{1}{2\pi i}\Lambda_t((t-z)^{-2}(t-z-h)^{-1}h).$$

Since $\Lambda$ belongs to $\mathcal{D}'_{L^p}$, it remains to prove that

$$g_{h,z}(\cdot) := (\cdot - z)^{-2}(\cdot - z - h)^{-1}h$$

tends to zero as $h \to 0$ in $\mathcal{D}_{L^q}$, where $\frac{1}{p} + \frac{1}{q} = 1$. The same computation as in the proof of Theorem 5.6.1 gives us

$$\frac{\partial^\alpha}{\partial t^\alpha}g_{h,z}(t) = (-1)^\alpha \alpha! h \sum_{\nu=0}^{\alpha}(\nu+1)(t-z)^{-(\nu+2)}(t-z-h)^{-(\alpha+1-\nu)}.$$

Putting $z = x + iy$ and $h = \xi + i\eta$ we get

$$\int_{\mathbb{R}} |(t-z)^{-(\nu+2)}(t-z-h)^{-(\alpha+1-\nu)}|^q\, dt$$

$$\leq \int_{\mathbb{R}}\left((t-x)^2 + d^2\right)^{-\frac{(\nu+2)q}{2}}\left((t-x-\xi)^2 + d^2\right)^{-\frac{(\alpha+1-\nu)q}{2}}\, dt =: A_\nu,$$

where $d = y - \delta$ if $y > 0$ and $d = |y + \delta|$ if $y < 0$. Then for $|h| < \delta < |y|$, by the Hölder inequality we get

$$A_\nu \leq \left(\int_{\mathbb{R}}\left((t-x)^2 + d^2\right)^{-\frac{(\nu+2)q^2}{2}}\, dt\right)^{\frac{1}{q}}\left(\int_{\mathbb{R}}\left((t-x-\xi)^2 + d^2\right)^{-\frac{(\alpha+1-\nu)pq}{2}}\, dt\right)^{\frac{1}{p}},$$

where $\frac{1}{p} + \frac{1}{q} = 1$ and $\nu = 0, 1, 2, \ldots$. Note that $\frac{(\nu+2)q^2}{2} > 1$ and $\frac{(\alpha+1-\nu)pq}{2} > \frac{1}{2}$ for $\nu = 0, 1, 2, \ldots \alpha$. Therefore the integrals

$$\int_{\mathbb{R}}\left((t-x)^2 + d^2\right)^{-\frac{(\nu+2)q^2}{2}}\, dt =: B_\nu$$

and

$$\int_{\mathbb{R}}\left((t-x-\xi)^2 + d^2\right)^{-\frac{(\alpha+1-\nu)pq}{2}}\, dt =: C_\nu$$

are finite and $A_\nu \leq B_\nu C_\nu$. Finally, we have

$$\|\partial^\alpha g_{h,z}(\cdot)\|_{L^q} \leq \alpha! |h| \sum_{\nu=0}^{\alpha}(\nu+1)B_\nu^{\frac{1}{q^2}} C_\nu^{\frac{1}{pq}}.$$

This implies that $\|\partial^\alpha g_{h,z}(\cdot)\|_{L^q}$ tends to zero as $h \to 0$. Thus, we have shown that

$$\frac{d}{dz} C\Lambda(z) = \frac{1}{2\pi i} \Lambda \left( \frac{1}{(\cdot - z)^2} \right).$$

The equality $\frac{1}{2\pi i} \Lambda \left( \frac{1}{(\cdot - z)^2} \right) = C\Lambda^{(1)}(z)$ is evident. This finishes the proof.    $\square$

We shall now show that each $\Lambda$ in $\mathcal{D}'_{L^p}$, $1 < p < \infty$ may be reproduced by means of its Cauchy transform. In order to get this result we shall need the following

LEMMA 5.7.1. *If* $\Lambda \in \mathcal{D}'_{L^p}$, $1 < p < \infty$, $g \in \mathcal{D}_{L^1}$ *and* $\int_{\mathbb{R}^n} g = 1$, *then*

$$\lim_{\epsilon \to 0} \int_{\mathbb{R}^n} \Lambda_{g_\epsilon}(x) \varphi(x) \, dx = \Lambda(\varphi)$$

*for* $\varphi \in \mathcal{D}_{L^q}$, $\frac{1}{p} + \frac{1}{q} = 1$, *where* $g_\epsilon(x) = \epsilon^{-n} g \left( \frac{x}{\epsilon} \right)$ *and* $\Lambda_{g_\epsilon}(x) = \Lambda(g_\epsilon(x - \cdot))$.

PROOF. First of all we have to show that $\Lambda_{g_\epsilon}$ is in $\mathcal{D}_{L^p}$. In view of Theorem 5.4.1 we have

$$\Lambda_{g_\epsilon}(x) = \sum_{|\alpha| \leq m} (-1)^{|\alpha|} \int_{\mathbb{R}^n} \frac{\partial^{|\alpha|}}{\partial y^\alpha} g_\epsilon(x - y) f_\alpha(y) \, dy,$$

where $f_\alpha$ is some fixed element of $L^p$. By the Young theorem we infer that $\Lambda_{g_\epsilon}$ is in $\mathcal{D}_{L^p}$ for $\epsilon > 0$. From this, it follows that the following equalities

$$\int_{\mathbb{R}^n} \Lambda_{g_\epsilon}(x) \varphi(x) \, dx = \int_{\mathbb{R}^n} \varphi(x) \left( \sum_{|\alpha| \leq m} (-1)^{|\alpha|} \int_{\mathbb{R}^n} \frac{\partial^{|\alpha|}}{\partial y^\alpha} g_\epsilon(x - y) f_\alpha(y) \, dy \right) dx$$

$$= \sum_{|\alpha| \leq m} (-1)^{|\alpha|} \int_{\mathbb{R}^n} \varphi(x) \left( \int_{\mathbb{R}^n} \frac{\partial^{|\alpha|}}{\partial y^\alpha} g_\epsilon(x - y) f_\alpha(y) \, dy \right) dx$$

hold. The last integral can be written as follows

$$(-1)^{|\alpha|} \int_{\mathbb{R}^n} \varphi(x) \left( \frac{\partial^{|\alpha|}}{\partial x^\alpha} \int_{\mathbb{R}^n} g_\epsilon(x - y) f_\alpha(y) \, dy \right) dx.$$

The $\alpha$-fold integration by parts (see Lemma 1.12.1) shows that

$$\int_{\mathbb{R}^n} \varphi(x) \left( \frac{\partial^{|\alpha|}}{\partial x^\alpha} \int_{\mathbb{R}^n} g_\epsilon(x - y) f_\alpha(y) \, dy \right) dx$$

$$= (-1)^{|\alpha|} \int_{\mathbb{R}^n} \frac{\partial^{|\alpha|}}{\partial x^\alpha} \varphi(x) \left( \int_{\mathbb{R}^n} g_\epsilon(x - y) f_\alpha(y) \, dy \right) dx.$$

On the other hand we have

$$\Lambda(\varphi) = \sum_{|\alpha| \leq m} (-1)^{|\alpha|} \int_{\mathbb{R}^n} f_\alpha(x) \frac{\partial^{|\alpha|}}{\partial x^\alpha} \varphi(x) \, dx.$$

Hence we get the following equality

$$\int_{\mathbb{R}^n} \Lambda_{g_\epsilon}(x)\varphi(x)\,dx - \Lambda(\varphi)$$

$$= \sum_{|\alpha|\leq m} (-1)^{|\alpha|} \int_{\mathbb{R}^n} \left( \int_{\mathbb{R}^n} (g_\epsilon(x-y)f_\alpha(y)\,dy - f_\alpha(x) \right) \frac{\partial^{|\alpha|}}{\partial x^\alpha} \varphi(x)\,dx.$$

It is easy to check that

$$\int_{\mathbb{R}^n} g_\epsilon(x-y)f_\alpha(y)\,dy - f_\alpha(x) = \int_{\mathbb{R}^n} (f_\alpha(x-\epsilon y) - f_\alpha(x))g(y)\,dy.$$

This implies that

$$\left| \int_{\mathbb{R}^n} \Lambda_{g_\epsilon}(x)\varphi(x)dx - \Lambda(\varphi) \right|$$

$$\leq \sum_{|\alpha|\leq m} \left| \int_{\mathbb{R}^n} \left( \int_{\mathbb{R}^n} (f_\alpha(x-\epsilon y) - f_\alpha(x))g(y)\,dy \right) \frac{\partial^{|\alpha|}}{\partial x^\alpha} \varphi(x)\,dx \right|$$

$$\leq \sum_{|\alpha|\leq m} \left( \int_{\mathbb{R}^n} \left( \int_{\mathbb{R}^n} |f_\alpha(x-\epsilon y) - f_\alpha(x)|\,|g(y)|\,dy \right)^p dx \right)^{\frac{1}{p}} \left( \int_{\mathbb{R}^n} \left| \frac{\partial^{|\alpha|}}{\partial x^\alpha} \varphi(x) \right|^q dx \right)^{\frac{1}{q}}.$$

It remains to be proved that the expression

$$\left( \int_{\mathbb{R}^n} \left( \int_{\mathbb{R}^n} |f_\alpha(x-\epsilon y) - f_\alpha(x)|\,|g(y)|\,dy \right)^p dx \right)^{\frac{1}{p}}$$

tends to zero as $\epsilon \to 0$. Using Minkowski's inequality we get

$$\left( \int_{\mathbb{R}^n} \left( \int_{\mathbb{R}^n} |f_\alpha(x-\epsilon y) - f_\alpha(x)|\,|g(y)|\,dy \right)^p dx \right)^{\frac{1}{p}}$$

$$\leq \int_{\mathbb{R}^n} |g(y)| \left( \int_{\mathbb{R}^n} |f_\alpha(x-\epsilon y) - f_\alpha(x)|^p\,dx \right)^{\frac{1}{p}} dy.$$

We have to show that the expression on the right side of this inequality tends to zero as $\epsilon \to 0$. Indeed, $\|f(\cdot - \epsilon y) - f\|_{L^p} \leq 2\|f\|_{L^p}$ for $y \in \mathbb{R}^n$ and $\|f(\cdot - \epsilon y) - f\|_{L^p} \to 0$ as $\epsilon \to 0$ for each $y \in \mathbb{R}^n$. Hence, by the Lebesgue dominated convergence theorem we obtain the required fact. This finishes the proof of the lemma. □

Now, we are able to prove a reproduced theorem for $\Lambda \in \mathcal{D}'_{L^p}$.

THEOREM 5.7.2. *If $\Lambda \in \mathcal{D}'_{L^p}$, $1 < p < \infty$, then*

$$\lim_{\epsilon \to 0} \int_R (C\Lambda(x + i\epsilon) - C\Lambda(x - i\epsilon))\varphi(x)\, dx = \Lambda(\varphi)$$

*for each $\varphi \in \mathcal{D}_{L^q}$, where $\frac{1}{p} + \frac{1}{q} = 1$.*

PROOF. Let $\Lambda \in \mathcal{D}'_{L^p}$. We know that

$$C\Lambda(x + i\epsilon) - C\Lambda(x - i\epsilon) = \Lambda_t \left( \frac{1}{\pi} \frac{\epsilon}{(t - x)^2 + \epsilon^2} \right).$$

Now, we see that the theorem is an immediate consequence of Lemma 5.7.1 if we take $g(x) = \frac{1}{\pi(x^2+1)}$. $\qquad\square$

## 5.8. Cauchy transforms of some distributions

In this section we shall be concerned with finite parts of some divergent integrals. Let $\varphi$ be in $\mathcal{D}_{L^\infty}$. Put by definition

$$\frac{1}{(\cdot)^{2m}}(\varphi) := \int_{\mathbb{R}} x^{-2m} \left( \frac{\varphi(x) + \varphi(-x)}{2} \right.$$
$$\left. - \left( \varphi(0) + \frac{1}{2!}\varphi^{(2)}(0)x^2 + \cdots + \frac{1}{(2m-2)!}\varphi^{(2m-2)}(0)x^{2m-2} \right) \right) dx \quad (5.8.1)$$

and

$$\frac{1}{(\cdot)^{2m+1}}(\varphi) := \int_{\mathbb{R}} x^{-(2m+1)} \left( \frac{\varphi(x) - \varphi(-x)}{2} \right.$$
$$\left. - \left( \frac{1}{1!}\varphi'(0)x + \cdots + \frac{1}{(2m-1)!}\varphi^{(2m-1)}(0)x^{2m-1} \right) \right) dx \quad (5.8.2)$$

for $m = 1, 2, \ldots$. For $m = 0$ we have

$$\frac{1}{(\cdot)}(\varphi) = \int_{\mathbb{R}} \frac{\varphi(x) - \varphi(-x)}{2x}\, dx \quad (5.8.3)$$

(compare with Example 1.1.4). We shall now show that equalities (5.8.1) and (5.8.2) are meaningful for $\varphi \in \mathcal{D}_{L^\infty}$ if $m = 1, 2, \ldots$. We restrict ourselves to consideration of (5.8.1) only. Formally (5.8.1) can be written as follows

$$\frac{1}{(\cdot)^{2m}}(\varphi) = \int_{|x|<1} x^{-2m} \left( \frac{\varphi(x) + \varphi(-x)}{2} - \left( \varphi(0) + \frac{1}{2!}\varphi^{(2)}(0)x^2 \right. \right.$$
$$\left. \left. + \cdots + \frac{1}{(2m-2)!}\varphi^{(2m-2)}(0)x^{2m-2} \right) \right) dx$$
$$+ \int_{|x|\geq 1} x^{-2m} \frac{\varphi(x) + \varphi(-x)}{2}\, dx - \int_{|x|\geq 1} \sum_{\nu=0}^{m-1} \frac{\varphi^{(2\nu)}(0)}{(2\nu)!} x^{-2(m-\nu)}\, dx.$$

It is easy to verify that the above integrals exist for $\varphi \in \mathcal{D}_{L^\infty}$. Moreover, these integrals determine linear continuous forms on $\mathcal{D}_{L^\infty}$ with respect to $\beta$-convergence (see Section 5.1). Therefore $\frac{1}{(\cdot)^{2m}}$ is in $\mathcal{D}'_{L^1}$. It is easy to show that $\frac{1}{(\cdot)}$ belongs to $\mathcal{D}'_{L^p}$, $1 < p < \infty$. Now, we determine the Cauchy transform of $\frac{1}{(\cdot)}$. Putting $\varphi(x) = \frac{1}{2\pi i} \frac{1}{x-z}$, $\Im x \neq 0$ into (5.8.3) we obtain

$$C \frac{1}{(\cdot)}(z) = \frac{1}{2\pi i} \int_{\mathbb{R}} \frac{1}{x^2 - z^2} dx.$$

An easy computation shows that

$$\frac{1}{2\pi i} \int_{\mathbb{R}} \frac{1}{x^2 - z^2} dx = \begin{cases} \frac{1}{2z} & \text{if } \Im z > 0 \\ -\frac{1}{2z} & \text{if } \Im z < 0. \end{cases} \tag{5.8.4}$$

To determine the Cauchy transforms of $\frac{1}{(\cdot)^k}$ for $k > 1$ we shall need a recurrent formula

$$C \frac{1}{(\cdot)^{k+1}}(z) = \frac{1}{z} C \frac{1}{(\cdot)^k}(z), \tag{5.8.5}$$

where $k = 1, 2, \cdots$ and $\Im z \neq 0$. Formulas (5.8.1) and (5.8.2) for $\varphi(x) = \frac{1}{2\pi i} \frac{1}{x-z}$ can be written as follows

$$C \frac{1}{(\cdot)^{2m}}(z) = \frac{1}{2\pi i} \int_{\mathbb{R}} x^{-2m} \left( \frac{z}{x^2 - z^2} + \frac{1}{z} + \frac{x^2}{z^3} + \cdots + \frac{x^{2(m-1)}}{z^{2m-1}} \right) dx$$

and

$$C \frac{1}{(\cdot)^{2m+1}}(z) = \frac{1}{2\pi i} \int_{\mathbb{R}} x^{-(2m+1)} \left( \frac{x}{x^2 - z^2} + \frac{x}{z^2} + \frac{x^3}{z^4} + \cdots + \frac{x^{2m-1}}{z^{2m}} \right) dx.$$

Let us notice that

$$C \frac{1}{(\cdot)^{2m}}(z) = \frac{1}{2\pi i} \frac{1}{z} \int_{\mathbb{R}} x^{-2m} \left( \frac{z^2}{x^2 - z^2} + 1 + \frac{x^2}{z^2} + \cdots + \frac{x^{2(m-1)}}{z^{2(m-1)}} \right) dx$$

$$= \frac{1}{2\pi i} \frac{1}{z} \int_{\mathbb{R}} x^{-2m+1} \left( \frac{x}{x^2 - z^2} + \frac{x}{z^2} + \cdots + \frac{x^{2(m-1)-1}}{z^{2(m-1)}} \right) dx$$

$$= \frac{1}{z} C \frac{1}{(\cdot)^{2m-1}}(z)$$

for $m = 1, 2; \ldots$ Analogously we obtain the following equality

$$C \frac{1}{(\cdot)^{2m+1}}(z) = \frac{1}{z} C \frac{1}{(\cdot)^{2m}}(z)$$

for $m = 1, 2 \ldots$ From this and (5.8.4) we get

$$C \frac{1}{(\cdot)^k}(z) = \begin{cases} \frac{1}{2z^k} & \text{if } \Im z > 0 \\ -\frac{1}{2z^k} & \text{if } \Im z < 0, \end{cases} \tag{5.8.6}$$

where $k = 1, 2, \ldots$.

This section will be finished by presenting the Sochocki–Plemelj formulas. We know that the Cauchy transform of Dirac's measure $\delta$ is given by formula

$$C\delta(z) = -\frac{1}{2\pi i}\frac{1}{z}, \quad z \neq 0.$$

Moreover, we have

$$\frac{d^k}{dz^k}C\delta(z) = C\delta^{(k)}(z) = \frac{k!(-1)^{k+1}}{2\pi i z^{k+1}}, \tag{5.8.7}$$

where $z \neq 0$. Hence we obtain

$$\frac{1}{2z^k} = \frac{\pi i(-1)^k}{(k-1)!}C\delta^{(k-1)}(z). \tag{5.8.8}$$

From this, it follows that

$$\frac{1}{2}\left(\frac{1}{(x+i\epsilon)^k} - \frac{1}{(x-i\epsilon)^k}\right) = \frac{\pi i(-1)^k}{(k-1)!}(C\delta^{(k-1)}(x+i\epsilon) - C\delta^{(k-1)}(x-i\epsilon)). \tag{5.8.9}$$

In view of (5.8.6) we obtain

$$\frac{1}{2}\left(\frac{1}{(x+i\epsilon)^k} + \frac{1}{(x-i\epsilon)^k}\right) = C\frac{1}{(\cdot)^k}(x+i\epsilon) - C\frac{1}{(\cdot)^k}(x-i\epsilon). \tag{5.8.10}$$

By virtue of (5.8.9) and (5.8.10) we get

$$\frac{1}{(x+i\epsilon)^k} = \frac{\pi i(-1)^k}{(k-1)!}(C\delta^{(k-1)}(x+i\epsilon) - C\delta^{(k-1)}(x-i\epsilon))$$
$$+ \left(C\frac{1}{(\cdot)^k}(x+i\epsilon) - C\frac{1}{(\cdot)^k}(x-i\epsilon)\right), \tag{5.8.11}$$

$$\frac{1}{(x-i\epsilon)^k} = -\frac{\pi i(-1)^k}{(k-1)!}(C\delta^{(k-1)}(x+i\epsilon) - C\delta^{(k-1)}(x-i\epsilon))$$
$$+ \left(C\frac{1}{(\cdot)^k}(x+i\epsilon) - C\frac{1}{(\cdot)^k}(x-i\epsilon)\right). \tag{5.8.12}$$

It was remarked previously that $\delta^{(k)}$ and $\frac{1}{(\cdot)^k}$ are in $\mathcal{D}'_{L^p}$, $1 < p < \infty$, $k = 1, 2, \ldots$. Therefore the expressions on the right side of (5.8.11) and (5.8.12) according to Theorem 5.7.2 tend to

$$\frac{\pi i(-1)^k}{(k-1)!}\delta^{(k-1)} + \frac{1}{(\cdot)^k} \quad \text{and} \quad -\frac{\pi i(-1)^k}{(k-1)!}\delta^{(k-1)} + \frac{1}{(\cdot)^k}$$

as $\epsilon \to 0$, respectively in $\mathcal{D}'_{L^p}$. One can remark that if $k \geq 2$, then the above convergence is in $\mathcal{D}'_{L^1}$. From this we infer that

$$\frac{1}{(\cdot + i\epsilon)^k} \to: \frac{1}{(\cdot + i0)^k} \quad \text{and} \quad \frac{1}{(\cdot - i\epsilon)^k} \to: \frac{1}{(\cdot - i0)^k}$$

as $\epsilon \to 0$ in $\mathcal{D}'_{L^p}$, $1 < p < \infty$ if $k = 1$ and in $\mathcal{D}'_{L^1}$ if $k \geq 2$. Finally we obtain the following Sochocki–Plemelj formulas

$$\frac{1}{(\cdot + \mathrm{i}0)^k} = \frac{\pi \mathrm{i}(-1)^k}{(k-1)!}\delta^{(k-1)} + \frac{1}{(\cdot)^k}, \tag{5.8.13}$$

$$\frac{1}{(\cdot - \mathrm{i}0)^k} = -\frac{\pi \mathrm{i}(-1)^k}{(k-1)!}\delta^{(k-1)} + \frac{1}{(\cdot)^k}, \tag{5.8.14}$$

(compare with (1.6.5) and (1.6.6)).

## CHAPTER 6

# TEMPERED DISTRIBUTIONS AND FOURIER TRANSFORMS

## 6.1. The spaces $S$ and $S'$

A space which very often occurs in applications is $S$, the space of infinitely differentiable functions defined on $\mathbb{R}^n$ which, together with their derivatives, approach zero more rapidly than any power of $\frac{1}{|x|}$ as $|x| \to \infty$ (for instance $\exp(-|x|^2)$). More precisely, we assume that $\varphi$ is in $S$ if there exists a constant $C_{\alpha\beta}$ such that

$$|x^\beta \partial^\alpha \varphi(x)| \le C_{\alpha\beta}$$

for $x \in \mathbb{R}^n$ and $\alpha, \beta \in \mathbb{N}^n$. One can introduce a family of norms in $S$ putting

$$\|\varphi\|_m := \max_{0 \le |\alpha+\beta| \le m} \sup_{x \in \mathbb{R}^n} |x^\beta \partial^\alpha \varphi(x)|. \tag{6.1.1}$$

DEFINITION 6.1.1. A sequence $(\varphi_\nu)$ is said to be convergent to $\varphi$ in $S$, if $\|\varphi_\nu - \varphi\|_m$ tends to zero as $\nu \to \infty$ for $m = 0, 1, 2, \ldots$

Obviously, $\varphi_\nu$ tends to $\varphi$ as $\nu \to \infty$ in $S$ if and only if $(\cdot)^\beta \partial^\alpha \varphi_\nu$ converges uniformly to $(\cdot)^\beta \partial^\alpha \varphi$ in $\mathbb{R}^n$ for all $\alpha, \beta \in \mathbb{N}^n$. We also note that if $\varphi_\nu, \varphi \in \mathcal{D}$ and $\varphi_\nu \to \varphi$ in $\mathcal{D}$, then clearly, $\varphi_\nu \to \varphi$ in $S$. In other words the convergence in $\mathcal{D}$ is stronger than convergence in $S$. One can show that $S$ is a complete space with respect to the family of norms (6.1.1). We see at once that if $\varphi$ is in $S$, then $(\cdot)^\beta \partial^\alpha \varphi$ is also in $L^1(\mathbb{R}^n)$ for all $\alpha, \beta \in \mathbb{N}^n$.

THEOREM 6.1.1. $\mathcal{D}$ is dense in $S$.

PROOF. Let $g_\nu$ be as in Section 1.12. We shall now show that for any $\varphi$ in $S$, the sequence $(g_\nu \varphi)$ converges to $\varphi$ in $S$. It is easy to check that

$$|\partial^\alpha g_\nu| \le M_\alpha$$

for $\nu = 1, 2 \ldots$. This involves the inclusion $(g_\nu \varphi)$ in $S$. Moreover, for fixed $\epsilon > 0$ there exists $\nu_0 \in \mathbb{N}$ such that

$$|x^\beta \partial^\alpha (g_\nu \varphi - \varphi)(x)| < \epsilon$$

for $x \notin B_{\nu_0 - 1}(0)$ and $\nu > \nu_0$. Further, we have

$$\partial^\alpha (g_\nu \varphi)(x) = (g_\nu \partial^\alpha \varphi)(x) = \partial^\alpha \varphi(x)$$

for $x \in B_{\nu-1}(0)$ and $\nu > \nu_0$ (see Theorem 5.1.3). Finally we obtain

$$|x^\beta \partial^\alpha (g_\nu \varphi - \varphi)(x)| < \epsilon$$

for $x \in \mathbb{R}^n$ and $\nu > \nu_0$. This statement finishes the proof. $\square$

DEFINITION 6.1.2. A linear form defined on $\mathcal{S}$ and continuous with respect to the convergence in this space (see Definition 6.1.1) is called a tempered distribution (exactly the restriction $\Lambda|_{\mathcal{D}}$ of $\Lambda$ into $\mathcal{D}$).

The set of all tempered distributions will be denoted by $\mathcal{S}'$.

DEFINITION 6.1.3. We say that $f \in \mathcal{O}_M$ if $f \in C^\infty(\mathbb{R}^n)$ and for each $\alpha \in \mathbb{N}^n$ there exist $M > 0$ and $k \in \mathbb{N}$ such that

$$\left| \frac{\partial^{|\alpha|}}{\partial x^\alpha} f(x) \right| \leq M(1 + |x|^2)^k.$$

REMARK 6.1.1. It is easy to check that if $\varphi \in \mathcal{S}$ and $f \in \mathcal{O}_M$, then $f\varphi$ is in $\mathcal{S}$. Moreover, if $\varphi_\nu \in \mathcal{S}$ and $\varphi_\nu$ converges to $\varphi$ in $\mathcal{S}$, then $f\varphi_\nu$ converges to $f\varphi$ in $\mathcal{S}$. This statement shows us that if $f$ is in $\mathcal{O}_M$ and $T$ is in $\mathcal{S}'$, then $fT$ is also in $\mathcal{S}'$ (see Section 1.7).

THEOREM 6.1.2. *To each $\Lambda$ in $\mathcal{S}'$ there correspond $m \in \mathbb{N}$ and complex Borel measures $\mu_{\alpha\beta}$ such that*

$$\Lambda(\varphi) = \sum_{0 \leq |\alpha+\beta| \leq m} (-1)^{|\alpha|} \int_{\mathbb{R}^n} (\cdot)^\beta \partial^\alpha \varphi \, d\mu_{\alpha\beta} \tag{6.1.2}$$

*for $\varphi \in \mathcal{S}$.*

PROOF. Since $\Lambda$ is a linear continuous form on $\mathcal{S}$, therefore there exist $m \in \mathbb{N}$ and $M > 0$ such that

$$|\Lambda(\varphi)| \leq M\|\varphi\|_m$$

for $\varphi \in \mathcal{S}$ (see Section 1.1). The rest of the proof runs as in Section 2.7. Let us consider the Cartesian product

$$C_0^N(\mathbb{R}^n) = \underset{0 \leq |\alpha+\beta| \leq m}{\text{X}} C_0(\mathbb{R}^n), \quad N = \sum_{0 \leq |\alpha+\beta| \leq m} 1.$$

We shall denote the elements of $C_0^N(\mathbb{R}^n)$ by $(\varphi_{\alpha\beta})$. The component of $(\varphi_{\alpha\beta})$ corresponding to subscript $\alpha\beta$ will be denoted by $\varphi_{\alpha\beta}$. Let us endow the Cartesian product $C_0^N(\mathbb{R}^n)$ with the norm

$$\|(\varphi_{\alpha\beta})\|_{C_0^N} := \max_{0 \leq |\alpha+\beta| \leq m} \sup_{x \in \mathbb{R}^n} |\varphi_{\alpha\beta}(x)|.$$

Set, by definition

$$\mathcal{S}^m := \{\varphi : \varphi \in C^\infty(\mathbb{R}^n), \ (\cdot)^\beta \partial^\alpha \varphi \in C_0(\mathbb{R}^n) \text{ for } |\alpha + \beta| \leq m\}.$$

Consider the mapping $\mathcal{K} : \mathcal{S}^m \to C_0^N(\mathbb{R}^n)$ defined in the following form

$$\mathcal{K}(\varphi) = ((-1)^{|\alpha|}(\cdot)^\beta \partial^\alpha \varphi).$$

Note that, $\mathcal{K}$ establishes the linear continuous injection from $\mathcal{S}^m$ into $C_0^N(\mathbb{R}^n)$. We shall now define a linear continuous form $\widetilde{\Lambda}$ on $\mathcal{K}(\mathcal{S}^m)$ taking

$$\widetilde{\Lambda}(\mathcal{K}(\varphi)) := \Lambda(\varphi).$$

Of course,

$$|\widetilde{\Lambda}(\mathcal{K}(\varphi))| = |\Lambda(\varphi)| \leq M\|\mathcal{K}(\varphi)\|_{C_0^N}.$$

By the Hahn–Banach theorem there exists a continuous extension of $\widetilde{\Lambda}$ from $\mathcal{K}(\mathcal{S}^m)$ onto the whole space $C_0^N(\mathbb{R}^n)$. By Lemma 2.6.1 and Theorem 2.7.2 we obtain (6.1.2). $\qquad\square$

## 6.2. Tensor product of tempered distributions

Let $\varphi(\cdot,\cdot)$ belong to $\mathcal{S}(\mathbb{R}^m \times \mathbb{R}^n)$. Let $S$ and $T$ be in $\mathcal{S}'(\mathbb{R}^m)$ and $\mathcal{S}'(\mathbb{R}^n)$, respectively. Put

$$(S_x \otimes T_y)\varphi(x,y) := S_x(T_y(\varphi(x,y))).$$

We shall now show that the right side of this equality is sensible for $\varphi \in \mathcal{S}(\mathbb{R}^m \times \mathbb{R}^n)$. In the first place we show that $\psi$ belongs to $\mathcal{S}(\mathbb{R}^m)$ if $\psi(x) := T_y(\varphi(x,y))$. Indeed, by Theorem 6.1.2 we have

$$\psi(x) = \sum_{0 \leq |\gamma+\delta| \leq l} (-1)^{|\gamma|} \int_{\mathbb{R}^n} y^\delta \frac{\partial^{|\gamma|}}{\partial y^\gamma} \varphi(x,y) \, d\mu_{\gamma\delta}(y)$$

for $\varphi(\cdot,\cdot) \in \mathcal{S}(\mathbb{R}^m \times \mathbb{R}^n)$. Put

$$\psi_{\gamma\delta}(x) := (-1)^{|\gamma|} \int_{\mathbb{R}^n} y^\delta \frac{\partial^{|\gamma|}}{\partial y^\gamma} \varphi(x,y) \, d\mu_{\gamma\delta}(y).$$

We begin by proving the continuity of $\psi_{\gamma\delta}$ in $\mathbb{R}^m$. Of course,

$$\psi_{\gamma\delta}(x+\xi) - \psi_{\gamma\delta}(x) = (-1)^{|\gamma|} \int_{\mathbb{R}^n} y^\delta \left( \frac{\partial^{|\gamma|}}{\partial y^\gamma} \varphi(x+\xi,y) - \frac{\partial^{|\gamma|}}{\partial y^\gamma} \varphi(x,y) \right) d\mu_{\gamma\delta}(y).$$

It is easily seen that

$$y^\delta \frac{\partial^{|\gamma|}}{\partial y^\gamma} \varphi(x+\xi,y) \to y^\delta \frac{\partial^{|\gamma|}}{\partial y^\gamma} \varphi(x,y)$$

uniformly with respect to $y \in \mathbb{R}^n$ as $\xi \to 0$ for arbitrary fixed $x \in \mathbb{R}^m$. Since the function $y^\beta \frac{\partial^{|\alpha|}}{\partial y^\beta} \varphi(x,y)$ is bounded in $\mathbb{R}^m \times \mathbb{R}^n$, therefore by the Lebesgue dominated convergence theorem we infer that $\psi_{\gamma\delta}$ is continuous. Further, we have

$$h^{-1}(\psi_{\gamma\delta}(x+he_j) - \psi_{\gamma\delta}(x)) = (-1)^{|\gamma|} \int_{\mathbb{R}^n} y^\delta \frac{\partial^{|\gamma|+1}}{\partial y^\gamma \partial x_j} \varphi(x+\theta he_j, y) \, d\mu_{\gamma\delta}(y),$$

where $0 < \theta < 1$. A passage to that limit similar to the above implies that

$$\frac{\partial}{\partial x_j} \psi_{\gamma\delta}(x) = (-1)^{|\gamma|} \int_{\mathbb{R}^n} y^\delta \frac{\partial^{|\gamma|+1}}{\partial y^\gamma \partial x_j} \varphi(x,y) \, d\mu_{\gamma\delta}(y).$$

We now proceed by induction on $\alpha$. Finally we get

$$\frac{\partial^{|\alpha|}}{\partial x^\alpha} \psi_{\gamma\delta}(x) = (-1)^{|\gamma|} \int_{\mathbb{R}^n} y^\delta \frac{\partial^{|\alpha+\gamma|}}{\partial x^\alpha \partial y^\gamma} \varphi(x,y) \, d\mu_{\gamma\delta}(y).$$

It remains to prove that the expression $x^\beta \frac{\partial^{|\alpha|}}{\partial x^\alpha} \psi_{\gamma\delta}(x)$ is bounded in $\mathbb{R}^m$. In fact, we have

$$\left| x^\beta \frac{\partial^{|\alpha|}}{\partial x^\alpha} \psi_{\gamma\delta}(x) \right| \leq \int_{\mathbb{R}^n} \sup_{(x,y) \in \mathbb{R}^m \times \mathbb{R}^n} \left| x^\beta y^\delta \frac{\partial^{|\alpha+\gamma|}}{\partial x^\alpha \partial y^\gamma} \varphi(x,y) \right| d|\mu_{\gamma\delta}|(y)$$

$$\leq M_{\alpha\beta\gamma\delta} |\mu_{\gamma\delta}|(\mathbb{R}^n),$$

where $|\mu_{\gamma\delta}|$ is the total variation of $\mu_{\gamma\delta}$. A simple verification shows that if $\varphi_\nu(\cdot,\cdot)$ tends to zero in $S(\mathbb{R}^m \times \mathbb{R}^n)$ then

$$(-1)^{|\gamma|} \int_{\mathbb{R}^n} y^\delta \frac{\partial^{|\gamma|}}{\partial x^\gamma} \varphi_\nu(\cdot,y) d\mu_{\gamma\delta}(y) \to 0$$

in $S(\mathbb{R}^n)$ as $\nu \to \infty$, as well. Using again Theorem 6.1.2 we obtain

$(S_x \otimes T_y)\varphi(x,y)$

$$= \sum_{|\alpha+\beta|\leq k} \sum_{|\gamma+\delta|\leq l} (-1)^{|\alpha+\gamma|} \int_{\mathbb{R}^m} \left( \int_{\mathbb{R}^n} x^\beta y^\delta \frac{\partial^{|\alpha+\gamma|}}{\partial x^\alpha \partial y^\gamma} \varphi(x,y) d\mu_{\gamma\delta}(y) \right) d\mu_{\alpha\beta}(x).$$

$$(6.2.1)$$

Of course, (6.2.1) is a linear continuous form on $S(\mathbb{R}^m \times \mathbb{R}^n)$.

DEFINITION 6.2.1. The linear continuous form on $S(\mathbb{R}^m \times \mathbb{R}^n)$ given by (6.2.1) is said to be the tensor product of $S$ and $T$.

COROLLARY 6.2.1. *Tensor product of tempered distributions is commutative.* *(See Section 5.3.)*

EXAMPLE 6.2.1. Let $1(x) = 1$ for each $x \in \mathbb{R}^m$. Assume that $T$ is in $S'(\mathbb{R}^n)$, then we have
$$(1_x \otimes T_y)\varphi(x,y) = (T_y \otimes 1_x)\varphi(x,y).$$
This equality can be written as follows

$$\int_{\mathbb{R}^m} T_y(\varphi(x,y))\,dx = T_y \left( \int_{\mathbb{R}^m} \varphi(x,y)\,dx \right).$$

## 6.3. Fourier transforms of integrable functions

In this section, we shall deal with a concept which is of basic importance in the applications of distributions to the theory of partial differential equations, in the theory of probability and many other branches of analysis. Let $f$ be in $L^1(\mathbb{R}^n)$. Put

$$\mathcal{F}f(\sigma) := \int_{\mathbb{R}^n} e^{i(x,\sigma)} f(x)\,dx,$$

$$(6.3.1)$$

where $(x,\sigma) = x_1\sigma_1 + \cdots + x_n\sigma_n$.

DEFINITION 6.3.1. The function $\mathcal{F}f$ is called the Fourier transform of $f$.

THEOREM 6.3.1. *If $f$ is in $L^1(\mathbb{R}^n)$, then $\mathcal{F}f$ is continuous and*

$$\|\mathcal{F}f\|_{L^\infty} \leq \|f\|_{L^1}.$$

$$(6.3.2)$$

PROOF. We see at once that $e^{i(x,\sigma+h)}$ tends pointwise to $e^{i(x,\sigma)}$ as $h \to 0$ and $|e^{i(x,\sigma+h)}f(x)| = |f(x)|$. Now, we can apply the Lebesgue dominated convergence theorem. The second part of the theorem is evident. $\qquad\square$

EXAMPLE 6.3.1. Let $f = \chi_{[-1,1]}$. An easy computation shows that

$$\mathcal{F}\chi_{[-1,1]}(\sigma) = 2\frac{\sin\sigma}{\sigma}.$$

It is easy to see that $\frac{\sin(\cdot)}{(\cdot)}$ does not belong to $L^1(\mathbb{R}^n)$.

EXAMPLE 6.3.2. Take $f(x) = e^{-\frac{x^2}{2}}$. Let us consider the contour

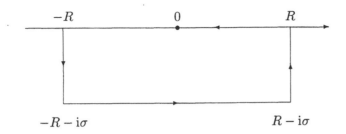

Now, $z \to e^{-\frac{z^2}{2}}$ is an entire function of the complex variable $z = x + iy$, and therefore its integral around the rectangle with vertices $-R$, $R$, $-R - i\sigma$ and $R - i\sigma$ is zero $(R > 0)$. It is easy to check that the integrals on the two vertical sides tend to zero as $R \to \infty$. Therefore

$$\int_{\mathbb{R}} e^{-\frac{1}{2}(x - i\sigma)^2} \, dx = \int_{\mathbb{R}} e^{-\frac{1}{2}x^2} \, dx.$$

On the other hand we have

$$\int_{\mathbb{R}} e^{ix\sigma} e^{-\frac{1}{2}x^2} \, dx = e^{-\frac{1}{2}\sigma^2} \int_{\mathbb{R}} e^{-\frac{1}{2}(x - i\sigma)^2} \, dx.$$

Since $\int_{\mathbb{R}} e^{-\frac{1}{2}x^2} \, dx = \sqrt{2\pi}$, therefore $\mathcal{F}\left(e^{-\frac{1}{2}(\cdot)^2}\right)(\sigma) = \sqrt{2\pi} e^{-\frac{1}{2}\sigma^2}$. In the same manner we can see that

$$\mathcal{F}\left(e^{-\frac{1}{2}|\cdot|^2}\right)(\sigma) = (2\pi)^{\frac{n}{2}} e^{-\frac{1}{2}|\sigma|^2}, \quad \sigma \in \mathbb{R}^n. \tag{6.3.3}$$

We also define

$$\mathcal{F}^{-1} f(\sigma) = \frac{1}{(2\pi)^n} \int_{\mathbb{R}^n} e^{-i(x,\sigma)} f(x) dx. \tag{6.3.4}$$

DEFINITION 6.3.2. The function $\mathcal{F}^{-1} f$ is said to be the inverse Fourier transform of $f$.

THEOREM 6.3.2 (Parseval). *If $f$ and $g$ are in $L^1(\mathbb{R}^n)$, then*

$$\int_{\mathbb{R}^n} \mathcal{F} f(\sigma) g(\sigma) \, d\sigma = \int_{\mathbb{R}^n} f(x) \mathcal{F} g(x) \, dx, \tag{6.3.5}$$

$$\int_{\mathbb{R}^n} \mathcal{F}^{-1} f(\sigma) g(\sigma) \, d\sigma = \int_{\mathbb{R}^n} f(x) \mathcal{F}^{-1} g(x) \, dx. \tag{6.3.6}$$

PROOF. The formulas (6.3.5) and (6.3.6) are an immediate consequence of Fubini's theorem. $\qquad\square$

We are now in a position to prove some Fourier transform inversion theorem.

THEOREM 6.3.3. *If $f$, $\mathcal{F} f$ are in $L^1(\mathbb{R}^n)$, then*

$$(\mathcal{F}^{-1} \circ \mathcal{F}) f = (\mathcal{F} \circ \mathcal{F}^{-1}) f = f. \tag{6.3.7}$$

PROOF. Note that if $f$ and $\mathcal{F}f$ belong to $L^1(\mathbb{R}^n)$, then $\mathcal{F}^{-1}f$ belongs to $L^1(\mathbb{R}^n)$, as well. Put by definition

$$G_{x\epsilon}(\sigma) := e^{-i(x,\sigma)-\frac{1}{2}\epsilon^2|\sigma|^2}.$$

We shall now apply the Fourier transformation with respect to $\sigma$. Thus, we have

$$g_{\epsilon x}(y) := \int_{\mathbb{R}^n} e^{i(y,\sigma)}e^{-i(x,\sigma)-\frac{1}{2}\epsilon^2|\sigma|^2}\,d\sigma = \int_{\mathbb{R}^n} e^{i(y-x,\sigma)}e^{-\frac{1}{2}\epsilon^2|\sigma|^2}\,d\sigma.$$

Putting $\omega = \epsilon\sigma$ we obtain

$$g_{\epsilon x}(y) = \epsilon^{-n}\int_{\mathbb{R}^n} e^{i(\frac{y-x}{\epsilon},\omega)}e^{-\frac{1}{2}|\omega|^2}\,d\omega.$$

By (6.3.3) we get

$$g_{\epsilon x}(y) = \epsilon^{-n}(2\pi)^{\frac{n}{2}}e^{-\frac{1}{2}\left|\frac{y-x}{\epsilon}\right|^2}.$$

It is easy to check that $(2\pi)^{-n}\int_{\mathbb{R}^n} g_{\epsilon x}(y)dy = 1$ for $x \in \mathbb{R}^n$. Because of (6.3.5) we have

$$\int_{\mathbb{R}^n} \mathcal{F}f(\sigma)G_{\epsilon x}(\sigma)\,d\sigma = \int_{\mathbb{R}^n} f(y)g_{\epsilon x}(y)\,dy. \tag{6.3.8}$$

Putting $\delta_\epsilon(x-y) = (2\pi)^{-n}g_{\epsilon x}(y)$ we obtain $\int_{\mathbb{R}^n} \delta_\epsilon(x-y)\,dy = 1$ for all $x \in \mathbb{R}^n$. Now, equality (6.3.8) can be written as follows

$$\int_{\mathbb{R}^n} \mathcal{F}f(\sigma)G_{\epsilon x}(\sigma)\,d\sigma = (2\pi)^n(f * \delta_\epsilon)(x).$$

By virtue of Lemma 1.2.1 we infer that $\int_{\mathbb{R}^n} \mathcal{F}f(\sigma)G_{\epsilon x}(\sigma)\,d\sigma$ tends to $(2\pi)^n f$ as $\epsilon \to 0$ in $L^1(\mathbb{R}^n)$. On the other hand we have

$$\int_{\mathbb{R}^n} \mathcal{F}f(\sigma)G_{\epsilon x}(\sigma)\,d\sigma = \int_{\mathbb{R}^n} \mathcal{F}f(\sigma)e^{-i(x,\sigma)-\frac{1}{2}\epsilon^2|\sigma|^2}\,d\sigma$$

$$= (2\pi)^n(2\pi)^{-n}\int_{\mathbb{R}^n} \mathcal{F}f(\sigma)e^{-i(x,\sigma)-\frac{1}{2}\epsilon^2|\sigma|^2}\,d\sigma.$$

Since $\mathcal{F}f$ is in $L^1(\mathbb{R}^n)$ and $\mathcal{F}f(\sigma)e^{-i(x,\sigma)-\frac{1}{2}\epsilon^2|\sigma|^2}$ tends to $\mathcal{F}f(\sigma)e^{-i(x,\sigma)}$ as $\epsilon \to 0$, by the Lebesgue dominated convergence theorem we deduce that

$$\int_{\mathbb{R}^n} \mathcal{F}f(\sigma)G_{\epsilon x}(\sigma)\,d\sigma \to (2\pi)^n(\mathcal{F}^{-1}\circ\mathcal{F})f(x)$$

for $x \in \mathbb{R}^n$. This implies that $(\mathcal{F}^{-1}\circ\mathcal{F})f(x) = f(x)$ almost everywhere in $\mathbb{R}^n$. In the same manner we can prove the second part of (6.3.7).                    □

## 6.4. Formal properties of Fourier transforms

From now on, we shall use alternately the notations $\mathcal{F}f$ or $\widehat{f}$ as the Fourier transform of $f$. We list the formal properties of $\mathcal{F}$ in the following

THEOREM 6.4.1. *If $f$ and $(\cdot)^{e_\nu} f$ are in $L^1(\mathbb{R}^n)$, then*

$$(\partial^{e_\nu} \widehat{f})(\sigma) = \widehat{(i\cdot)^{e_\nu} f}(\sigma). \tag{6.4.1}$$

*If $f$ and $\partial^{e_\nu} f$ are in $L^1(\mathbb{R}^n)$, then*

$$\widehat{\partial^{e_\nu} f} = (-i\sigma)^{e_\nu} \widehat{f}(\sigma). \tag{6.4.2}$$

*If $f$ is in $L^1(\mathbb{R}^n)$, then we have*

$$\widehat{\tau_x f}(\sigma) = e^{i(x,\sigma)} \widehat{f}(\sigma) \tag{6.4.3}$$

*for each $x \in \mathbb{R}^n$ (see Example 1.8.4).*
*If $f$ is in $L^1(\mathbb{R}^n)$, then we have*

$$\widehat{e^{i(x,\cdot)} f}(\sigma) = \tau_{-x} \widehat{f}(\sigma) \tag{6.4.4}$$

*for $x \in \mathbb{R}^n$.*
*If $f$ is in $L^1(\mathbb{R}^n)$ and $K : \mathbb{R}^n \to \mathbb{R}^n$ is an orthogonal transformation, then*

$$\widehat{f \circ K}(\sigma) = (\widehat{f} \circ K)(\sigma). \tag{6.4.5}$$

*If $f$ is in $L^1(\mathbb{R}^n)$ and $S_\lambda f(y) = \lambda^n f(\lambda y)$, $\lambda > 0$, then*

$$\widehat{(S_\lambda f}(\sigma) = \widehat{f}\left(\frac{1}{\lambda}\sigma\right). \tag{6.4.6}$$

*If $f$ and $g$ are in $L^1(\mathbb{R}^n)$, then*

$$\widehat{f * g}(\sigma) = \widehat{f}(\sigma)\widehat{g}(\sigma) \quad \text{(Borel formula).} \tag{6.4.7}$$

PROOF. Ad (6.4.1). Note that

$$(\widehat{f}(\sigma + he_\nu) - \widehat{f}(\sigma))h^{-1} = h^{-1} \int_{\mathbb{R}^n} (e^{i(x,\sigma+he_\nu)} - e^{i(x,\sigma)}) f(x)\, dx.$$

Since

$$e^{i(x,he_\nu)} - 1 = \int_0^h (ix)^{e_\nu} e^{i(x,te_\nu)}\, dt,$$

therefore $|(e^{i(x,he_\nu)} - 1)h^{-1}| \le |x_\nu|$ for $h \ne 0$. On the other hand $(e^{i(x,he_\nu)} - 1)h^{-1}$ tends to $(ix)^{e_\nu}$ as $h \to 0$. From this, by the Lebesgue dominated convergence theorem we obtain

$$\lim_{h \to 0} (\widehat{f}(\sigma + he_\nu) - \widehat{f}(\sigma))h^{-1} = \int_{\mathbb{R}^n} e^{i(x,\sigma)} (ix)^{e_\nu} f(x)\, dx.$$

This completes the proof of (6.4.1).

Ad (6.4.2). We can write $x$ and $\sigma$ in the following form $x = (x', x_\nu)$ and $\sigma = (\sigma', \sigma_\nu)$, where $x'$ and $\sigma'$ are in $\mathbb{R}^{n-1}$. Note that the following equalities

$$\int_{\mathbb{R}^n} e^{i(x,\sigma)} \partial^{e_\nu} f(x) \, dx = \int_{\mathbb{R}^{n-1}} e^{i(x',\sigma')} \left( \int_{\mathbb{R}} e^{i(x_\nu,\sigma_\nu)} \partial^{e_\nu} f(x) \, dx_\nu \right) dx'$$

$$= \int_{\mathbb{R}^{n-1}} e^{i(x',\sigma')} \left( \int_{\mathbb{R}} (-i\sigma)^{e_\nu} e^{i(x_\nu,\sigma_\nu)} f(x) \, dx_\nu \right) dx'$$

$$= (-i\sigma)^{e_\nu} \widehat{f}(\sigma)$$

are true. This proves (6.4.2).

The formulas (6.4.3) and (6.4.4) can be obtained by direct substitution in the definition formula of Fourier transform.

To prove (6.4.5) we compute the integral

$$\int_{\mathbb{R}^n} e^{i(y,\sigma)} (f \circ K)(y) \, dy = \int_{\mathbb{R}^n} e^{i(y,\sigma)} f(Ky) \, dy.$$

Taking $\omega = Ky$ we get $y = K^T \omega$, where $K^T$ denotes the transpose matrix of $K$. We obtain

$$\widehat{f \circ K}(\sigma) = \int_{\mathbb{R}^n} e^{i(K^T\omega,\sigma)} f(\omega) \, d\omega = \int_{\mathbb{R}^n} e^{i(\omega,K\sigma)} f(\omega) \, d\omega = (\widehat{f} \circ K)(\sigma).$$

Formula (6.4.6) can be determined by a simple verification. The proof of (6.4.7) is an application of Fubini's theorem. $\qquad\square$

## 6.5. Fourier transforms of functions in $\mathcal{S}$

It is easily seen that the following formulas

$$\partial^\alpha \widehat{\varphi}(\sigma) = \widehat{(i\cdot)^\alpha \varphi}(\sigma) \tag{6.5.1}$$

$$\widehat{\partial^\alpha \varphi}(\sigma) = (-i\sigma)^\alpha \widehat{\varphi}(\sigma) \tag{6.5.2}$$

are true if $\varphi$ is in $\mathcal{S}$. We are able to prove the following

THEOREM 6.5.1. *If $\varphi$ is in $\mathcal{S}$, then $\mathcal{F}\varphi$ is in $\mathcal{S}$, too. Moreover, the mapping*

$$\varphi \to \mathcal{F}\varphi \tag{6.5.3}$$

*is continuous.*

PROOF. On account of (6.5.1) and (6.5.2) we have

$$\sigma^\alpha \partial^\beta \widehat{\varphi}(\sigma) = \sigma^\alpha \widehat{(i\cdot)^\beta \varphi}(\sigma) = i^{|\beta+\alpha|} \widehat{\partial^\alpha((\cdot)^\beta \varphi)}(\sigma). \tag{6.5.4}$$

Since $\partial^\alpha((\cdot)^\beta \varphi)$ is in $\mathcal{S}$, therefore by Theorem 6.3.1 we conclude that $\partial^\beta \widehat{\varphi}$ is continuous and $\sigma^\alpha \partial^\beta \widehat{\varphi}$ is bounded in $\mathbb{R}^n$. This means that $\widehat{\varphi}$ belongs to $\mathcal{S}$. It remains to prove the continuity of (6.5.3). In accordance with (6.5.4) for $\beta \geq \alpha$ we have

$$\sup_{\sigma \in \mathbb{R}^n} |\sigma^\alpha \partial^\beta \widehat{\varphi}(\sigma)| = \sup_{\sigma \in \mathbb{R}^n} |\widehat{\partial^\alpha ((\cdot)^\beta \varphi)}(\sigma)| \leq \int_{\mathbb{R}^n} |\partial^\alpha ((\cdot)^\beta \varphi)(x)| \, dx$$

$$= \int_{\mathbb{R}^n} \left| \sum_{0 \leq \mu \leq \alpha} \binom{\alpha}{\mu} \prod_{k=1}^n \beta_k (\beta_k - 1) \cdots (\beta_k - \mu_k + 1) x^{\beta - \mu} \partial^{\alpha - \mu} \varphi(x) \right| dx$$

$$\leq K \sum_{0 \leq \mu \leq \alpha} \binom{\alpha}{\mu} \prod_{k=1}^n \beta_k (\beta_k - 1) \cdots (\beta_k - \mu_k + 1) \sup_{x \in \mathbb{R}^n} ((1 + |x|^2)^m |x^{\beta - \mu} \partial^{\alpha - \mu} \varphi(x)|,$$

where $K = \int_{\mathbb{R}^n} (1 + |x|^2)^{-m} dx$, $m > \frac{n}{2}$. Generally, one can show that there exists $M$ such that

$$\|(\cdot)^\alpha \partial^\beta \widehat{\varphi}\|_{L^\infty} \leq M \sum_{0 \leq \mu \leq \alpha} \|P_\mu \partial^{\alpha - \mu} \varphi\|_{L^\infty},$$

where $P_\mu$ is a polynomial. This inequality implies the continuity of (6.5.3) and the proof is complete. $\qquad\square$

REMARK 6.5.1. Formula (6.3.7) is true if $f$ is in $\mathcal{S}$.

We are able to prove a stronger theorem than Theorem 6.3.1.

THEOREM 6.5.2 (Riemann–Lebesgue lemma). *If $f$ is in $L^1(\mathbb{R}^n)$, then $\mathcal{F}f$ is in* $C_0(\mathbb{R}^n)$.

PROOF. In view of Theorem 1.5.1 there exists a sequence $(f_\nu)$, $f_\nu \in \mathcal{D}$ convergent in $L^1(\mathbb{R}^n)$ to $f$. In accordance with Theorem 6.5.1, $\mathcal{F}f_\nu \in \mathcal{S}$. Taking into account Theorem 6.3.1 we infer that $\mathcal{F}f_\nu$ converges uniformly in $\mathbb{R}^n$ to $\mathcal{F}f$. For given $\epsilon > 0$ there exists $\nu_0 \in \mathbb{N}$ such that

$$|\mathcal{F}f_\nu(\sigma) - \mathcal{F}f(\sigma)| < \frac{\epsilon}{2}$$

for $\sigma \in \mathbb{R}^n$, $\nu > \nu_0$. Since $\mathcal{F}f_\nu$ is in $\mathcal{S}$ therefore $|\mathcal{F}_\nu f(\sigma)| < \frac{\epsilon}{2}$ for $|\sigma| > \sigma_0$. This implies that $|\mathcal{F}f(\sigma)| < \epsilon$ if $|\sigma| > \sigma_0$ Thus, the proof is finished. $\qquad\square$

## 6.6. Fourier transforms of functions in $L^2(\mathbb{R}^n)$

Let $f$ be in $L^1(\mathbb{R}^n)$. Put $\tilde{f}(x) := \overline{f(-x)}$. We shall now show that

$$\widehat{\tilde{f}} = \overline{\widehat{f}}. \tag{6.6.1}$$

Note that $\overline{ab} = \overline{a}\,\overline{b}$ for $a, b \in \mathbb{C}$. Let $f$ be a simple integrable function. It is easily seen that

$$\overline{\widehat{f}}(\sigma) = \overline{\int_{\mathbb{R}^n} e^{i(z,\sigma)} f(z) \, dz} = \int_{\mathbb{R}^n} e^{-i(z,\sigma)} \overline{f(z)} \, dz = \int_{\mathbb{R}^n} e^{i(y,\sigma)} \overline{f(-y)} \, dy = \widehat{\tilde{f}}(\sigma).$$

In the general case, we get (6.6.1) by limit proceeding in $L^1(\mathbb{R}^n)$.

We are now in a position to prove the following

THEOREM 6.6.1. *If $\varphi$ is in $\mathcal{S}$, then*

$$\|\mathcal{F}\varphi\|_{L^2} = (2\pi)^{\frac{n}{2}} \|\varphi\|_{L^2} \tag{6.6.2}$$

*and*

$$\|\mathcal{F}^{-1}\varphi\|_{L^2} = (2\pi)^{-\frac{n}{2}} \|\varphi\|_{L^2}. \tag{6.6.3}$$

PROOF. Let $\varphi$ belong to $\mathcal{S}$. Then the following equalities

$$\|\varphi\|_{L^2}^2 = \int\limits_{\mathbb{R}^n} |\varphi(x)|^2\, dx = \int\limits_{\mathbb{R}^n} \varphi(x)\overline{\varphi(x)}\, dx = \int\limits_{\mathbb{R}^n} \varphi(x)\tilde{\varphi}(-x)\, dx = (\varphi * \tilde{\varphi})(0)$$

hold. On the other hand we have

$$(2\pi)^{-n} \int\limits_{\mathbb{R}^n} \widehat{\varphi * \tilde{\varphi}}(x)\, dx = (2\pi)^{-n} \int\limits_{\mathbb{R}^n} \widehat{\varphi * \tilde{\varphi}}(x) e^{-i(x,0)}\, dx$$

$$= (\mathcal{F}^{-1} \circ \mathcal{F})(\varphi * \tilde{\varphi})(0) = (\varphi * \tilde{\varphi})(0).$$

From this it follows that

$$\|\varphi\|_{L^2}^2 = (2\pi)^{-n} \int\limits_{\mathbb{R}^n} \widehat{\varphi * \tilde{\varphi}}(x)\, dx.$$

By (6.4.7) and (6.6.1) we obtain (6.6.2). Putting $\mathcal{F}^{-1}\varphi$ into (6.6.2) we get (6.6.3). The proof is complete. □

Let $f$ be in $L^2(\mathbb{R}^n)$. In accordance with Theorem 1.5.1 there exists a sequence $(\varphi_\nu)$, $\varphi_\nu \in \mathcal{D}$ such that $\varphi_\nu$ tends to $f$ as $\nu \to \infty$ in $L^2(\mathbb{R}^n)$. Because of (6.6.2) the sequence $(\mathcal{F}\varphi_\nu)$ is a Cauchy sequence in $L^2(\mathbb{R}^n)$. Suppose now that $\varphi_\nu \in \mathcal{S}$ and $\varphi_\nu$ converges to $f$ in $L^2(\mathbb{R}^n)$. Then $\mathcal{F}\varphi_\nu$ converges to some function $g$ also in $L^2(\mathbb{R}^n)$. After the above preliminaries we shall introduce the following definition.

DEFINITION 6.6.1. The function $g$ above defined is called the Fourier (Plancherel) transform of $f$.

To verify the consistency of this definition we have to show that if the sequences $(\varphi_\nu^1)$ and $(\varphi_\nu^2)$ converge to $f$, then the sequence $(\mathcal{F}\varphi_\nu^1 - \mathcal{F}\varphi_\nu^2)$ converges to zero in $L^2(\mathbb{R}^n)$. This fact is a consequence of (6.6.2). Similarly we can define the inverse Fourier transform of any function $f$ belonging to $L^2(\mathbb{R}^n)$.

DEFINITION 6.6.2. Suppose that $\varphi_\nu \in \mathcal{S}$ and the sequence $(\varphi_\nu)$ converges to $f$ in $L^2(\mathbb{R}^n)$. Then we take

$$\mathcal{F}^{-1}f = \lim_{\nu \to \infty} \mathcal{F}^{-1}\varphi_\nu$$

To verify the consistency of this definition we apply (6.6.3).

THEOREM 6.6.2 (Plancherel). *If $f$ is in $L^2(\mathbb{R}^n)$, then*

$$(\mathcal{F}^{-1} \circ \mathcal{F})f = (\mathcal{F} \circ \mathcal{F}^{-1})f = f, \tag{6.6.4}$$

$$\|\mathcal{F}f\|_{L^2} = (2\pi)^{\frac{n}{2}}\|f\|_{L^2} \text{ and } \|\mathcal{F}^{-1}f\|_{L^2} = (2\pi)^{-\frac{n}{2}}\|f\|_{L^2}. \tag{6.6.5}$$

PROOF. Let $\varphi_\nu$ be in $\mathcal{S}$. Assume that the $\varphi_\nu$ converge to $f$ in $L^2(\mathbb{R}^n)$. Then the sequences $((\mathcal{F}^{-1} \circ \mathcal{F})\varphi_\nu)$ and $((\mathcal{F} \circ \mathcal{F}^{-1})\varphi_\nu)$ converge to $(\mathcal{F}^{-1} \circ \mathcal{F})f$ and $(\mathcal{F} \circ \mathcal{F}^{-1})f$ in $L^2(\mathbb{R}^n)$, respectively. Since $(\mathcal{F}^{-1} \circ \mathcal{F})\varphi_\nu = (\mathcal{F} \circ \mathcal{F}^{-1}))\varphi_\nu = \varphi_\nu$, $\nu = 1, 2, \ldots$, therefore (6.6.4) is true. Equality (6.6.5) is an immediate consequence of the continuity of the function $L^2 \ni \varphi \to \|\varphi\|_{L^2}$. This finishes the proof. □

THEOREM 6.6.3 (Parseval). *If $f$ and $g$ are in $L^2(\mathbb{R}^n)$, then*

$$\int\limits_{\mathbb{R}^n} \mathcal{F}fg = \int\limits_{\mathbb{R}^n} f\mathcal{F}g. \tag{6.6.6}$$

PROOF. Let $\varphi_\nu$, $\psi_\nu$, $\nu = 1, 2, \ldots$ be in $\mathcal{S}$. Assume that the sequences $(\varphi_\nu)$ and $(\psi_\nu)$ converge to $f$ and $g$, respectively in $L^2(\mathbb{R}^n)$. By Theorem 6.3.2 we have

$$\int_{\mathbb{R}^n} \widehat{\varphi_\nu} \psi_\nu = \int_{\mathbb{R}^n} \varphi_\nu \widehat{\psi_\nu}.$$

What is left to show is that $\int_{\mathbb{R}^n} \widehat{\varphi_\nu} \psi_\nu$ and $\int_{\mathbb{R}^n} \varphi_\nu \widehat{\psi_\nu}$ converge to $\int_{\mathbb{R}^n} \mathcal{F}f g$ and $\int_{\mathbb{R}^n} f \mathcal{F}g$ respectively. In fact, we have

$$\left| \int_{\mathbb{R}^n} \mathcal{F}f g - \int_{\mathbb{R}^n} \widehat{\varphi_\nu} \psi_\nu \right| \leq \|\mathcal{F}f\|_{L^2} \|g - \psi_\nu\|_{L^2} + M \|\mathcal{F}f - \widehat{\varphi_\nu}\|_{L^2},$$

where $M = \sup_{\nu \in \mathbb{N}} \|\psi_\nu\|_{L^2}$. The rest of the proof is evident. $\qquad \square$

We shall now show that if $f$ belongs both to $L^1(\mathbb{R}^n)$ and $L^2(\mathbb{R}^n)$, then Definitions 6.3.1 and 6.6.1 are equivalent. Indeed, assume that $f$ is in $L^2(\mathbb{R}^n) \cap L^1(\mathbb{R}^n)$. Further, let $\mathcal{F}f$ denote the Fourier transform of $f$ in the sense of Definition 6.6.1. Let us regard for a moment $f$ as an element of $L^1(\mathbb{R}^n)$, then by Theorem 6.3.2

$$\int_{\mathbb{R}^n} \mathcal{F}f \varphi = \int_{\mathbb{R}^n} f \widehat{\varphi}$$

for each $\varphi$ in $\mathcal{S}$. In this case $\mathcal{F}f$ denotes the Fourier integral of $f$. On the other hand we can regard $f$ as an element of $L^2(\mathbb{R}^n)$, then the above equality may be understood in the sense of Theorem 6.6.3. From this we conclude that $\mathcal{F}f$ is the Fourier integral of $f$.

Let $\chi_{[-\nu,\nu]}$ denote the characteristic function of the set

$$\{x : x \in \mathbb{R}^n, \ -\nu \leq x_j \leq \nu, \ j = 1, \ldots, n\}.$$

The next theorem is very important for applications.

THEOREM 6.6.4. If $f$ is in $L^2(\mathbb{R}^n)$, then $\widehat{\chi_{[-\nu,\nu]}f}$ converges to $\mathcal{F}f$ in $L^2(\mathbb{R}^n)$.

PROOF. Of course, $\chi_{[-\nu,\nu]}f$ belongs to $L^1(\mathbb{R}^n)$, therefore

$$\widehat{\chi_{[-\nu,\nu]}f}(\sigma) = \int_{\mathbb{R}^n} e^{i(x,\sigma)} \chi_{[-\nu,\nu]}(x) f(x) \, dx.$$

Note that $(\widehat{\chi_{[-\nu,\nu]}f})$ is a Cauchy sequence in $L^2(\mathbb{R}^n)$. Put

$$f^* = \lim_{\nu \to \infty} \widehat{\chi_{[-\nu,\nu]}f}.$$

We have to show that $f^* = \mathcal{F}f$. Let $\varphi \in \mathcal{S}$, then by Theorem 6.3.2 we have

$$\int_{\mathbb{R}^n} \widehat{\chi_{[-\nu,\nu]}f} \varphi = \int_{\mathbb{R}^n} \chi_{[-\nu,\nu]} f \widehat{\varphi}.$$

A passage to limit implies that

$$\int_{\mathbb{R}^n} f^* \varphi = \int_{\mathbb{R}^n} f \widehat{\varphi}. \tag{6.6.7}$$

On the other hand by Theorem 6.6.3 we have

$$\int_{\mathbb{R}^n} f\hat{\varphi} = \int_{\mathbb{R}^n} \mathcal{F}f\varphi. \tag{6.6.8}$$

From (6.6.7) and (6.6.8) it follows that $f^* = \mathcal{F}f$. Thus, the proof is finished. $\quad\square$

To determine Fourier transforms of some functions we often use the following

LEMMA 6.6.1 (Jordan). *Let $f$ be a continuous function in the upper or lower half plane. If $\sigma > 0$ and $\lim_{\substack{|z|\to\infty \\ \Im z > 0}} f(z) = 0$ or $\sigma < 0$ and $\lim_{\substack{|z|\to\infty \\ \Im z < 0}} f(z) = 0$, then*

$$\lim_{r\to\infty} \int_{K_r^+} f(z)e^{i\sigma z}\, dz = 0 \quad \text{or} \quad \lim_{r\to\infty} \int_{K_r^-} f(z)e^{i\sigma z}\, dz = 0,$$

*where $K_r^+ = \{z : z \in \mathbb{C},\ |z| = r,\ \Im z \geq 0\}$ and $K_r^- = \{z : z \in \mathbb{C},\ |z| = r,\ \Im z \leq 0\}$.*

PROOF. We only shall prove the case, when $f$ is defined in the upper half plane. Put $M(r) := \sup_{t\in[0,\pi]} |f(re^{it})|$. Then we have

$$\left| \int_{K_r^+} f(z)e^{i\sigma z}\, dz \right| \leq M(r)r \int_0^\pi e^{-\sigma r \sin t}\, dt.$$

Since $\lim_{r\to\infty} M(r) = 0$, therefore what is left to show is that the integral $r\int_0^\pi e^{-\sigma r \sin t}\, dt$ is bounded for fixed $\sigma > 0$. An easy computation shows that

$$r \int_0^\pi e^{-\sigma r \sin t}\, dt = 2r \int_0^{\frac{\pi}{2}} e^{-\sigma r \sin t}\, dt.$$

Note that $\sin t \geq \frac{2t}{\pi}$ for $0 \leq t \leq \frac{\pi}{2}$. Therefore $e^{-\sigma r \sin t} \leq e^{-\frac{2\sigma r t}{\pi}}$ for $0 \leq t \leq \frac{\pi}{2}$. Hence, we obtain

$$2r \int_0^{\frac{\pi}{2}} e^{-\frac{2\sigma r t}{\pi}}\, dt \leq \frac{\pi}{\sigma}. \qquad\qquad \square$$

EXAMPLE 6.6.1. Let $f(x) = \frac{x}{1+x^2}$. We shall now compute the Fourier transform of $f$. Note that for $\sigma > 0$ we have

$$\frac{1}{2\pi i} \int_{-r}^{r} e^{i\sigma x} \frac{x}{1+x^2}\, dx + \frac{1}{2\pi i} \int_{K_r^+} e^{i\sigma z} \frac{z}{1+z^2}\, dz = \frac{1}{2} e^{-\sigma}$$

if $r > 1$. Hence, by Lemma 6.6.1 and Theorem 6.6.4 we obtain

$$\mathcal{F}f(\sigma) = \pi i e^{-\sigma}$$

for $\sigma > 0$. Similarly, one can show that

$$\mathcal{F}f(\sigma) = -\pi i e^{\sigma}$$

if $\sigma < 0$.

EXAMPLE 6.6.2. Put $f(x) := \frac{1}{2\pi i} \frac{1}{(x-z)}$, $\Im z \neq 0$. First we suppose that $\Im z > 0$. Then we have

$$\frac{1}{2\pi i} \int_{-r}^{r} e^{i\sigma x} \frac{1}{x-z}\, dx + \frac{1}{2\pi i} \int_{K_r^+} e^{i\sigma \xi} \frac{1}{\xi - z}\, d\xi = e^{i\sigma z}$$

for $\sigma > 0$ and

$$\frac{1}{2\pi i} \int_{-r}^{r} e^{i\sigma x} \frac{1}{x-z}\, dx + \frac{1}{2\pi i} \int_{K_r^-} e^{i\sigma \xi} \frac{1}{\xi - z}\, d\xi = 0$$

for $\sigma < 0$, where $|z| < r$. From this by Theorem 6.6.4 and Lemma 6.6.1 we obtain

$$\mathcal{F}\left(\frac{1}{2\pi i} \frac{1}{(\cdot - z)}\right)(\sigma) = H(\sigma)e^{i\sigma z}$$

if $\Im z > 0$, where $H$ is Heaviside step function. Similarly, assuming that $\Im z < 0$ we get $\mathcal{F}\left(\frac{1}{2\pi i} \frac{1}{(\cdot - z)}\right)(\sigma) = -H(-\sigma)e^{i\sigma z}$.

## 6.7. Fourier transforms of the Hermite functions

First we recall the definition of the Hermite functions.

DEFINITION 6.7.1. The function

$$H_\nu(x) = (-1)^\nu e^{x^2} (e^{-x^2})^{(\nu)}, \quad \nu = 0, 1 \ldots \tag{6.7.1}$$

is called the Hermite polynomial.

Carrying out differentiation in (6.7.1) we find $H_0(x) = 1$, $H_1(x) = 2x$, $H_2(x) = 4x^2 - 2, \ldots$ A simple verification shows that

$$\int_{\mathbb{R}} e^{-x^2} H_\mu(x) H_\nu(x)\, dx = \begin{cases} 0 & \text{if } \mu \neq \nu \\ 2^k k! \sqrt{\pi} & \text{if } \mu = \nu = k. \end{cases}$$

Set

$$h_k(x) := (2^k k! \sqrt{\pi})^{-\frac{1}{2}} e^{-\frac{x^2}{2}} H_k(x).$$

It is known that the functions $h_k$, $k = 0, 1, 2..$ constitute a complete orthonormal system in $L^2(\mathbb{R})$. It is called the Hermite system. The following differential operator

$$A = e^{\frac{x^2}{2}} \frac{d}{dx} e^{-\frac{x^2}{2}} = \frac{d}{dx} - x$$

will be useful to determine the Fourier transform of $h_k$. Namely, we shall show that

$$Ah_k = -(2(k+1))^{\frac{1}{2}} h_{k+1}, \quad k = 0, 1, 2, \ldots \tag{6.7.2}$$

Indeed, it is easy to check that

$$h'_k = (-1)^k (2^k k! \sqrt{\pi})^{-\frac{1}{2}} \left(x e^{\frac{x^2}{2}} (e^{-x^2})^{(k)} + e^{\frac{x^2}{2}} (e^{-x^2})^{(k+1)}\right).$$

On the other hand we have

$$-xh_k(x) = (-1)^{k+1} (2^k k! \sqrt{\pi})^{-\frac{1}{2}} \left(x e^{\frac{x^2}{2}} (e^{-x^2})^{(k)}\right).$$

Finally we have (6.7.2). In accordance with (6.4.1) and (6.4.2) we have

$$\mathcal{F}(A\varphi) = iA\mathcal{F}\varphi, \quad \varphi \in \mathcal{S}. \tag{6.7.3}$$

We are now in a position to prove the following

THEOREM 6.7.1.
$$\mathcal{F}h_k = 2\pi i^k h_k, \quad k = 0, 1, 2, \ldots \tag{6.7.4}$$

PROOF. Of course, $h_0 = e^{\frac{-x^2}{2}}$ and $\mathcal{F}h_0 = 2\pi h_0$. Assume now, that (6.7.4) is true for $n = k$, $k \geq 1$. By virtue of (6.7.2) we get $h_{k+1} = -(2(k+1))^{-\frac{1}{2}}Ah_k$. In accordance with (6.7.3)

$$\mathcal{F}h_{k+1} = -(2(k+1))^{-\frac{1}{2}}iA\mathcal{F}h_k.$$

From this, by the induction assumption we get

$$\mathcal{F}h_{k+1} = -(2(k+1))^{-\frac{1}{2}}i^{k+1}2\pi Ah_k.$$

Using (6.7.2) again we obtain $\mathcal{F}h_{k+1} = 2\pi i^{k+1}h_{k+1}$. This statement finishes the proof. $\square$

## 6.8. Fourier transforms of tempered distributions

We recall now that if $f$ is in $L^1(\mathbb{R}^n)$ or $L^2(\mathbb{R}^n)$, then we have

$$\int_{\mathbb{R}^n} \mathcal{F}f\varphi = \int_{\mathbb{R}^n} f\mathcal{F}\varphi \tag{6.8.1}$$

for $\varphi \in \mathcal{S}$. This leads us to

DEFINITION 6.8.1. The Fourier transform $\mathcal{F}T$ $(\widehat{T})$ of the distribution $T \in \mathcal{S}'(\mathbb{R}^n)$ is the image of $T$ under the transpose of the map $\mathcal{F} : \mathcal{S}(\mathbb{R}^n) \to \mathcal{S}(\mathbb{R}^n)$; i.e.

$$\mathcal{F}T(\varphi) = T(\mathcal{F}\varphi) \tag{6.8.2}$$

for $\varphi \in \mathcal{S}$. Similarly, one can define the inverse Fourier transform $\mathcal{F}^{-1}T$ of the tempered distribution $T$ taking

$$\mathcal{F}^{-1}(T(\varphi)) = T(\mathcal{F}^{-1}(\varphi)). \tag{6.8.3}$$

It is easily seen that if $T$ is in $L^1(\mathbb{R}^n)$ or $L^2(\mathbb{R}^n)$, then the classical definitions of $\mathcal{F}T$ and Definition 6.8.1 are equivalent (see Theorems 6.3.2 and 6.6.3).

THEOREM 6.8.1. *For a tempered distribution $T$ the following equalities*
$$(\mathcal{F}^{-1} \circ \mathcal{F})T = (\mathcal{F} \circ \mathcal{F}^{-1})T = T \tag{6.8.4}$$
*hold.*

PROOF. To prove this statement we only need to observe that

$$(\mathcal{F}^{-1} \circ \mathcal{F})T(\varphi) = (\mathcal{F}^{-1}(\mathcal{F}T))(\varphi) = \mathcal{F}T(\mathcal{F}^{-1}(\varphi)) = T(\mathcal{F} \circ \mathcal{F}^{-1})(\varphi) = T(\varphi).$$

The second part of (6.8.4) can be proved in the same way. $\square$

THEOREM 6.8.2. *The Fourier transformation $\mathcal{F} : \mathcal{S}'(\mathbb{R}^n) \to \mathcal{S}'(\mathbb{R}^n)$ is continuous.*

PROOF. Let $T_\nu$ and $T$ be in $\mathcal{S}'(\mathbb{R}^n)$, $T_\nu$ tends to $T$ as $\nu \to \infty$ in $\mathcal{S}'(\mathbb{R}^n)$. This means that $T_\nu(\varphi) \to T(\varphi)$ as $\nu \to \infty$ for each $\varphi \in \mathcal{S}$. Therefore

$$\widehat{T_\nu}(\varphi) = T_\nu(\widehat{\varphi}) \to T(\widehat{\varphi}) = \widehat{T}(\varphi)$$

as $\nu \to \infty$. This finishes the proof. $\square$

EXAMPLE 6.8.1. We shall now determine the Fourier transform $\mathcal{F}\frac{1}{(\cdot)}$ of $\frac{1}{(\cdot)}$. We know that $\frac{1}{(\cdot)}$ is in $\mathcal{D}'_{L^2}(\mathbb{R})$. Let $C\frac{1}{(\cdot)}$ be the Cauchy transform of $\frac{1}{(\cdot)}$. By virtue of Theorem 5.7.2 we infer that

$$C\frac{1}{(\cdot)}(\cdot+i\epsilon) - C\frac{1}{(\cdot)}(\cdot-i\epsilon) \to \frac{1}{(\cdot)}$$

as $\epsilon \to 0^+$ in $\mathcal{D}'_{L^2}(\mathbb{R})$. Therefore $C\frac{1}{(\cdot)}(\cdot+i\epsilon) - C\frac{1}{(\cdot)}(\cdot-i\epsilon)$ tends to $\frac{1}{(\cdot)}$ as $\epsilon \to 0^+$ in $\mathcal{S}'(\mathbb{R})$. It remains to determine the Fourier transforms of the functions $C\frac{1}{(\cdot)}(\cdot+i\epsilon)$ and $C\frac{1}{(\cdot)}(\cdot-i\epsilon)$. In accordance with (5.8.6), $C\frac{1}{(\cdot)}(z) = \frac{1}{2z}$ if $\Im z > 0$ and $C\frac{1}{(\cdot)}(z) = -\frac{1}{2z}$ if $\Im z < 0$. Using Theorem 6.6.4 and Lemma 6.6.1 we obtain

$$\frac{1}{2}\int_{\mathbb{R}} e^{ix\sigma}\frac{1}{x+i\epsilon}\,dx = \begin{cases}0 & \text{for } \sigma > 0 \\ -\pi i e^{\epsilon\sigma} & \text{for } \sigma < 0\end{cases}$$

and

$$\frac{1}{2}\int_{\mathbb{R}} e^{ix\sigma}\frac{1}{x-i\epsilon}\,dx = \begin{cases}\pi i e^{-\epsilon\sigma} & \text{for } \sigma > 0 \\ 0 & \text{for } \sigma < 0.\end{cases}$$

To obtain the Fourier transform $\mathcal{F}\frac{1}{(\cdot)}$ of $\frac{1}{(\cdot)}$ we compute

$$\lim_{\epsilon\to0^+}\frac{1}{2}\left(\mathcal{F}\frac{1}{(\cdot+i\epsilon)}(\sigma)+\mathcal{F}\frac{1}{(\cdot-i\epsilon)}(\sigma)\right) = \begin{cases}\pi i & \text{if } \sigma > 0 \\ -\pi i & \text{if } \sigma < 0.\end{cases}$$

Finally we have

$$\mathcal{F}\frac{1}{(\cdot)}(\sigma) = \pi i\,\mathrm{sgn}\,\sigma.$$

## 6.9. Formal properties of Fourier transforms of tempered distributions

We shall now show that formulas (6.4.1), (6.4.2), (6.4.3) and (6.4.4) are meaningful for tempered distributions.

THEOREM 6.9.1. *If $T$ is a tempered distribution, then the following equalities hold:*

$$D^{e_\nu}\widehat{T} = \widehat{(i\cdot)^{e_\nu}T}, \tag{6.9.1}$$

$$\widehat{D^{e_\nu}T} = (-i\cdot)^{e_\nu}\widehat{T}, \tag{6.9.2}$$

$$\widehat{\tau_x T} = e^{i(x,\cdot)}\widehat{T} \quad \text{for each} \quad x \in \mathbb{R}^n, \tag{6.9.3}$$

$$\widehat{e^{i(x,\cdot)}T} = \tau_{-x}\widehat{T} \quad \text{for each} \quad x \in \mathbb{R}^n. \tag{6.9.4}$$

PROOF. Formulas (6.9.1) and (6.9.2) are an immediate consequence of (6.4.1) and (6.4.2), respectively. Indeed, we have

$$D^{e_\nu}\widehat{T}(\varphi) = -\widehat{T}(\partial^{e_\nu}\varphi) = -T(\widehat{\partial^{e_\nu}\varphi}) = -T((-i\cdot)^{e_\nu}\widehat{\varphi})$$

$$= (i\cdot)^{e_\nu}T(\widehat{\varphi}) = \widehat{(i\cdot)^{e_\nu}T}(\varphi),$$

$$\widehat{D^{e_\nu}T}(\varphi) = D^{e_\nu}T(\widehat{\varphi}) = -T(\partial^{e_\nu}\widehat{\varphi}) = -T(\widehat{(i\cdot)^{e_\nu}\varphi}) = (-i\cdot)^{e_\nu}\widehat{T}(\varphi).$$

We shall now show that equalities (6.9.3) and (6.9.4) are also true. In accordance with Example 1.8.4, (6.4.3) and (6.4.4) we obtain

$$\widehat{\tau_x T}(\varphi) = \tau_x T(\widehat{\varphi}) = T(\tau_{-x}\widehat{\varphi}) = T(\widehat{e^{i(x,\cdot)}\varphi}) = e^{i(x,\cdot)}\widehat{T}(\varphi)$$

and

$$\widehat{e^{i(x,\cdot)}T}(\varphi) = (e^{i(x,\cdot)}T)(\widehat{\varphi}) = T(e^{i(x,\cdot)}\widehat{\varphi}) = T(\widehat{\tau_x\varphi}) = \widehat{T}(\tau_x\varphi) = \tau_{-x}\widehat{T}(\varphi).$$

The consequence of this reason is easily seen because of Remark 6.1.1.          □

## 6.10. Fourier transforms of integrable distributions

Put

$$E_\lambda(x) := (2\pi)^{-\frac{n}{2}}\lambda^{-n}e^{-\frac{1}{2}|\frac{x}{\lambda}|^2}$$

(compare Section 4.3). Let $T$ be in $\mathcal{D}'_{L^1}$, then

$$T_\lambda(x) = (T * E_\lambda)(x) = T(E_\lambda(x - \cdot))$$

is a regularization of $T$ (see Section 5.2). The first theorem of this section gives us some generalization of the Riemann–Lebesgue lemma.

THEOREM 6.10.1. *If $T$ is in $\mathcal{D}'_{L^1}$, then*

$$\mathcal{F}T(\sigma) = T(e^{i(\cdot,\sigma)}). \tag{6.10.1}$$

*Moreover, $\mathcal{F}T$ is a continuous function and bounded by a polynomial.*

PROOF. It is known that $E_1$ is in $\mathcal{D}_{L^1}$ and $\int_{\mathbb{R}^n} E_1 = 1$. In view of Theorem 5.3.2, $T_\lambda$ tends to $T$ in $\mathcal{D}'_{L^1}$ as $\lambda \to 0$. By the continuity of the Fourier transformation in $\mathcal{S}'(\mathbb{R}^n)$ it follows that $\mathcal{F}T_\lambda$ tends to $\mathcal{F}T$ in $\mathcal{S}'(\mathbb{R}^n)$ as $\lambda \to 0$. We shall now show that $\mathcal{F}T_\lambda(\sigma)$ converges to $T(e^{i(\cdot,\sigma)})$ as $\lambda \to 0$ at each point $\sigma \in \mathbb{R}^n$. In accordance with Theorem 5.1.4 we have

$$\mathcal{F}T_\lambda(\sigma) = \int_{\mathbb{R}^n} e^{i(x,\sigma)} \left( \sum_{|\alpha|\leq m} (-1)^{|\alpha|} \int_{\mathbb{R}^n} \frac{\partial^{|\alpha|}}{\partial y^\alpha} E_\lambda(x - y) d\mu_\alpha(y) \right) dx$$

$$= \sum_{|\alpha|\leq m} (-1)^{|\alpha|} \int_{\mathbb{R}^n} \left( \int_{\mathbb{R}^n} e^{i(x,\sigma)}(-1)^{|\alpha|}\frac{\partial^{|\alpha|}}{\partial x^\alpha} E_\lambda(x - y) dx \right) d\mu_\alpha(y)$$

$$= \sum_{|\alpha|\leq m} (-1)^{|\alpha|} \int_{\mathbb{R}^n} (i\sigma)^\alpha \left( \int_{\mathbb{R}^n} e^{i(x,\sigma)} E_\lambda(x - y) dx \right) d\mu_\alpha(y)$$

$$= \sum_{|\alpha|\leq m} (-1)^{|\alpha|} \int_{\mathbb{R}^n} (i\sigma)^\alpha e^{i(y,\sigma)} \left( \int_{\mathbb{R}^n} e^{i(\omega,\sigma)} E_\lambda(\omega) d\omega \right) d\mu_\alpha(y)$$

$$= \sum_{|\alpha|\leq m} (-1)^{|\alpha|} \int_{\mathbb{R}^n} (i\sigma)^\alpha e^{i(y,\sigma)} e^{-\frac{1}{2}|\lambda\sigma|^2} d\mu_\alpha(y).$$

Therefore $\mathcal{F}T_\lambda(\sigma)$ converges to

$$\sum_{|\alpha|\leq m} (-1)^{|\alpha|} \int_{\mathbb{R}^n} (i\sigma)^\alpha e^{i(y,\sigma)} d\mu_\alpha(y) = \sum_{|\alpha|\leq m} (-1)^{|\alpha|} \int_{\mathbb{R}^n} \frac{\partial^{|\alpha|}}{\partial y^\alpha} e^{i(y,\sigma)} d\mu_\alpha(y)$$

$$= T(e^{i(\cdot,\sigma)}).$$

Note that

$$|\mathcal{F}T_\lambda(\sigma)| \leq \sum_{|\alpha|\leq m} |\sigma_1|^{\alpha_1} \dots |\sigma_n|^{\alpha_n} |\mu_\alpha|(\mathbb{R}^n).$$

We shall now show that

$$|\mathcal{F}T_\lambda(\sigma)\varphi(\sigma)| \le M_\varphi(1+|\sigma|^2)^{-k},$$

where $k > \frac{n}{2}$. To prove this, it suffices to remark that

$$|\mathcal{F}T_\lambda(\sigma)\varphi(\sigma)| \le |\varphi(\sigma)|(1+|\sigma|^2)^k(2\pi)^{-\frac{n}{2}} \sum_{|\alpha|\le m} |\mu_\alpha|(\mathbb{R}^n)\,|\sigma_1|^{\alpha_1} \ldots |\sigma_n|^{\alpha_n}(1+|\sigma|^2)^{-k}.$$

Then we can set

$$M_\varphi := \sup_{\sigma \in \mathbb{R}^n} |\varphi(\sigma)|(1+|\sigma|^2)^k(2\pi)^{-\frac{n}{2}} \sum_{|\alpha|\le m} |\mu_\alpha|(\mathbb{R}^n)\,|\sigma_1|^{\alpha_1} \ldots |\sigma_n|^{\alpha_n}.$$

After these statements, by the Lebesgue dominated convergence theorem we infer that

$$\int_{\mathbb{R}^n} \mathcal{F}T_\lambda(\sigma)\varphi(\sigma)\,d\sigma \to \int_{\mathbb{R}^n} T(e^{i(\cdot,\sigma)})\varphi(\sigma)\,d\sigma$$

as $\lambda \to 0$ for each $\varphi \in \mathcal{S}$. The proof will be completed by showing that $\mathcal{F}T$ is continuous. The continuity of $\mathcal{F}T$ is easy to see from the representation formula

$$\mathcal{F}T(\sigma) = \sum_{|\alpha|\le m} (-1)^{|\alpha|} \int_{\mathbb{R}^n} \frac{\partial^{|\alpha|}}{\partial y^\alpha} e^{i(y,\sigma)} d\mu_\alpha(y). \qquad \square$$

THEOREM 6.10.2. *If $S$ and $T$ belong to $\mathcal{D}'_{L^1(\mathbb{R}^n)}$, then $S * T$ is in $\mathcal{D}'_{L^1(\mathbb{R}^n)}$ and the Borel formula*

$$\mathcal{F}(S * T)(\sigma) = \mathcal{F}S(\sigma)\,\mathcal{F}T(\sigma) \qquad (6.10.2)$$

*holds.*

PROOF. Indeed, we have

$$\mathcal{F}(S * T)(\sigma) = (S_x \otimes T_y)(e^{i(x+y,\sigma)}) = S_x(e^{i(x,\sigma)})S_y(e^{i(y,\sigma)})$$
$$= \mathcal{F}S(\sigma)\,\mathcal{F}T(\sigma).$$

This finishes the proof. $\qquad \square$

REMARK 6.10.1. Let $S$ and $T$ be in $\mathcal{D}'_{L^1(\mathbb{R}^m)}$ and $\mathcal{D}'_{L^1(\mathbb{R}^n)}$, respectively and let $\mathcal{F}S$ and $\mathcal{F}T$ be their Fourier transforms. Then a simple verification shows that

$$\mathcal{F}(S \otimes T)(\xi,\eta) = \mathcal{F}S(\xi)\,\mathcal{F}T(\eta).$$

In other words, the Fourier transform of the tensor product of integrable distributions is the tensor product of their Fourier transforms.

EXAMPLE 6.10.1. It is clear that $\mathcal{F}\delta = 1$ for $\delta \in \mathcal{D}'(\mathbb{R})$. Therefore for $\delta_\sigma \in \mathcal{D}'(\mathbb{R}^n)$ we have

$$\mathcal{F}\delta_\sigma = \mathcal{F}(\delta_{\sigma_1} \otimes \cdots \otimes \delta_{\sigma_n}) = 1,$$

where $\sigma = (\sigma_1, \ldots, \sigma_n)$.

## 6.11. Fourier transforms of square integrable distributions

A distribution belonging to $\mathcal{D}'_{L^2}$ from henceforth will be called a square integrable distribution. Denote by $\mathcal{S}'_0$ the vector space spanned by the set of all functions $(\cdot)^\nu f$, where $f$ is in $L^2(\mathbb{R}^n)$.

DEFINITION 6.11.1. A function $f$ belonging to $\mathcal{S}'_0$ will be called slowly increasing or tempered function.

THEOREM 6.11.1. $\mathcal{F}\mathcal{D}'_{L^2} \subset \mathcal{S}'_0$ and $\mathcal{F}\mathcal{S}'_0 \subset \mathcal{D}'_{L^2}$.

PROOF. Let $T \in \mathcal{D}'_{L^2}$, then in accordance with Theorem 5.4.1 we have

$$T(\varphi) = \sum_{|\alpha| \leq m} (-1)^{|\alpha|} \int_{\mathbb{R}^n} f_\alpha \partial^\alpha \varphi = \left( \sum_{|\alpha| \leq m} D^\alpha f_\alpha \right)(\varphi)$$

where $D^\alpha f_\alpha$ is the $\alpha$-distributional derivative of $f_\alpha$ belonging to $L^2(\mathbb{R}^n)$ and $\varphi$ is in $\mathcal{S}$. According to Definition 6.8.1

$$\mathcal{F}T(\varphi) = \left( \sum_{|\alpha| \leq m} D^\alpha f_\alpha \right)(\widehat{\varphi}) = \sum_{|\alpha| \leq m} (-1)^{|\alpha|} \int_{\mathbb{R}^n} f_\alpha \partial^\alpha \widehat{\varphi} = \sum_{|\alpha| \leq m} (-1)^{|\alpha|} \int_{\mathbb{R}^n} f_\alpha \widehat{(i\cdot)^\alpha \varphi}.$$

By Theorem 6.6.3 we have

$$\mathcal{F}T(\varphi) = \sum_{|\alpha| \leq m} (-1)^{|\alpha|} \int_{\mathbb{R}^n} (ix)^\alpha \mathcal{F}f_\alpha(x)\varphi(x)\, dx.$$

This implies that $\mathcal{F}T$ is in $\mathcal{S}'_0$. Conversely, if $T$ belongs to $\mathcal{S}'_0$, then

$$T(\varphi) = \sum_{|\alpha| \leq m} \int_{\mathbb{R}^n} x^\alpha f_\alpha(x)\varphi(x)\, dx,$$

where $f_\alpha \in L^2(\mathbb{R}^n)$ and $\varphi \in \mathcal{S}$. Therefore

$$\mathcal{F}T(\varphi) = \sum_{|\alpha| \leq m} \int_{\mathbb{R}^n} x^\alpha f_\alpha(x)\widehat{\varphi}(x)\, dx = \sum_{|\alpha| \leq m} \int_{\mathbb{R}^n} i^{|\alpha|}(-ix)^\alpha f_\alpha(x)\widehat{\varphi}(x)\, dx$$

$$= \sum_{|\alpha| \leq m} \int_{\mathbb{R}^n} i^{|\alpha|} f_\alpha(x)\widehat{\partial^\alpha \varphi}(x)\, dx.$$

Using again Theorem 6.6.3 we get

$$\mathcal{F}T(\varphi) = \sum_{|\alpha| \leq m} (-1)^{|\alpha|} \int_{\mathbb{R}^n} (-i)^{|\alpha|} \mathcal{F}f_\alpha(x)\partial^\alpha \varphi(x)\, dx.$$

Since $f_\alpha$ is in $L^2(\mathbb{R}^n)$, therefore $\mathcal{F}T$ is in $\mathcal{D}'_{L^2}$. This statement finishes the proof. $\square$

The next theorem is very useful generalization of Theorem 6.10.2.

THEOREM 6.11.2. If $S$ and $T$ are in $\mathcal{D}'_{L^2}$, then

$$\widehat{S * T} = \widehat{S}\, \widehat{T}. \tag{6.11.1}$$

PROOF. Let us begin by proving (6.11.1) for $S = f$, $T = g$, where $f$ and $g$ are in $L^2(\mathbb{R}^n)$. In accordance with Definition 6.8.1 we have

$$\mathcal{F}(f * g)(\varphi) = \int_{\mathbb{R}^n} (f * g)(x)\widehat{\varphi}(x)\,dx$$

for $\varphi \in \mathcal{S}$. First of all we shall show that $f * g$ is in $C_0(\mathbb{R}^n)$. Indeed, let $f_\nu$ and $g_\nu$ be in $\mathcal{S}$ for each $\nu \in \mathbb{N}$. Assume that the sequences $(f_\nu)$ and $(g_\nu)$ converge to $f$ and $g$, respectively in $L^2(\mathbb{R}^n)$. Of course, $f_\nu * g_\nu$ is in $\mathcal{S}$, therefore by Young's inequality we get

$$\|f * g - f_\nu * g_\nu\|_{L^\infty} \le \|f_\nu - f\|_{L^2}\|g\|_{L^2} + M\|g_\nu - g\|_{L^2},$$

where $M = \sup_{\nu \in \mathbb{N}}\{\|f_\nu\|_{L^2}\}$. This implies that $f * g \in C_0(\mathbb{R}^n)$. Further, $\int_{\mathbb{R}^n}(f_\nu * g_\nu)\widehat{\varphi}$ tends to $\int_{\mathbb{R}^n}(f * g)\widehat{\varphi}$ as $\nu \to \infty$. By (6.4.7) and Theorem 6.6.3 we get

$$\int_{\mathbb{R}^n} (f_\nu * g_\nu)\widehat{\varphi} = \int_{\mathbb{R}^n} \widehat{f_\nu}\widehat{g_\nu}\varphi.$$

On the other hand we have

$$\|\widehat{f_\nu}\widehat{g_\nu} - \mathcal{F}f\mathcal{F}g\|_{L^1} \le \|\widehat{f_\nu} - \mathcal{F}f\|_{L^2}\|\mathcal{F}g\|_{L^2} + (2\pi)^{\frac{n}{2}}M\|\widehat{g_\nu} - \mathcal{F}g\|_{L^2}.$$

In accordance with Theorem 6.6.2, $\widehat{f_\nu}\widehat{g_\nu}$ tends to $\mathcal{F}f\mathcal{F}g$ in $L^1(\mathbb{R}^n)$. This implies that $\int_{\mathbb{R}^n}\widehat{f_\nu}\widehat{g_\nu}\varphi$ tends to $\int_{\mathbb{R}^n}\widehat{f}\widehat{g}\varphi$ as $\nu \to \infty$. Finally we get

$$\int_{\mathbb{R}^n} (f * g)\widehat{\varphi} = \int_{\mathbb{R}^n} \widehat{f}\widehat{g}\varphi$$

for $f, g$ in $L^2(\mathbb{R}^n)$. This means that (6.11.1) is true if $S$ and $T$ are in $L^2(\mathbb{R}^n)$.

We now turn to the general case. By (5.5.5) we have

$$(S * T)(\varphi) = \sum_{|\alpha| \le k}\sum_{|\beta| \le l}(-1)^{|\alpha+\beta|}\int_{\mathbb{R}^n}\int_{\mathbb{R}^n} f_\alpha(x)g_\beta(y)\partial^{\alpha+\beta}\varphi(x+y)\,dx\,dy, \quad (6.11.2)$$

where $f_\alpha$ and $g_\beta$ are in $L^2(\mathbb{R}^n)$ and $\varphi \in \mathcal{S}$. Using the Fubini theorem we obtain

$$\int_{\mathbb{R}^n}\int_{\mathbb{R}^n} f_\alpha(x)g_\beta(y)\varphi(x+y)\,dx\,dy = \int_{\mathbb{R}^n} f_\alpha(x)\left(\int_{\mathbb{R}^n} g_\beta(y)\varphi(x+y)\,dy\right)\,dx$$

$$= \int_{\mathbb{R}^n} f_\alpha(x)\left(\int_{\mathbb{R}^n} g_\beta(\omega - x)\varphi(\omega)\,d\omega\right)\,dx = \int_{\mathbb{R}^n} \varphi(\omega)\left(\int_{\mathbb{R}^n} g_\beta(\omega - x)f_\alpha(x)\,dx\right)\,d\omega$$

$$= \int_{\mathbb{R}^n} (f_\alpha * g_\beta)(\omega)\varphi(\omega)\,d\omega.$$

Therefore equality (6.11.2) can be written as follows

$$(S * T)\varphi = \sum_{|\alpha| \le k}\sum_{|\beta| \le l}(-1)^{|\alpha+\beta|}\int_{\mathbb{R}^n}(f_\alpha * g_\beta)(x)\partial^{\alpha+\beta}\varphi(x)\,dx. \quad (6.11.3)$$

From this we get

$$\mathcal{F}(S * T)(\varphi) = (S * T)(\widehat{\varphi}) = \sum_{|\alpha| \leq k} \sum_{|\beta| \leq l} (-1)^{|\alpha+\beta|} \int_{\mathbb{R}^n} (f_\alpha * g_\beta) \partial^{\alpha+\beta} \widehat{\varphi}(x) \, dx$$

$$= \sum_{|\alpha| \leq k} \sum_{|\beta| \leq l} \int_{\mathbb{R}^n} (f_\alpha * g_\beta)(\widehat{-i \cdot})^{\alpha+\beta} \varphi(x) \, dx.$$

Taking into account the first part of the proof we obtain

$$\mathcal{F}(S * T)(\varphi) = \sum_{|\alpha| \leq k} \sum_{|\beta| \leq l} \int_{\mathbb{R}^n} \widehat{f_\alpha}(\sigma) \widehat{g_\beta}(\sigma) (-i\sigma)^{\alpha+\beta} \varphi(\sigma) \, d\sigma$$

$$= \sum_{|\alpha| \leq k} \sum_{|\beta| \leq l} \int_{\mathbb{R}^n} \widehat{D^\alpha f_\alpha}(\sigma) \widehat{D^\beta g_\beta}(\sigma) \varphi(\sigma) \, d\sigma.$$

Since $S = \sum_{|\alpha| \leq k} D^\alpha f_\alpha$ and $T = \sum_{|\beta| \leq l} D^\beta g_\beta$, therefore (6.11.1) generally holds. This finishes the proof.    □

A simple verification shows that

$$\mathcal{F}^{-1}(\varphi * \psi)(\sigma) = (2\pi)^n (\mathcal{F}^{-1}\varphi)(\sigma)(\mathcal{F}^{-1}\psi)(\sigma)$$

if $\varphi$ and $\psi$ are in $S$. Similar arguments applying to the inverse Fourier transform give us the following formula

$$\mathcal{F}^{-1}(S * T) = (2\pi)^n \mathcal{F}^{-1} S \, \mathcal{F}^{-1} T \qquad (6.11.4)$$

for $S$ and $T$ in $\mathcal{D}'_{L^2}$.

Thus, we obtain the following

THEOREM 6.11.3. *If $S$ and $T$ are tempered distributions in $L^2_{loc}(\mathbb{R}^n)$, then*

$$S \cdot T = \frac{1}{(2\pi)^n} \mathcal{F}^{-1}(\mathcal{F}S * \mathcal{F}T). \qquad (6.11.5)$$

## 6.12. Determining Fourier transforms of square integrable distributions

Formula (6.10.1) cannot be used to determining the Fourier transform $\mathcal{F}T$ of $T$ if $T$ is in $\mathcal{D}'_{L^2}$, because $e^{i(\cdot, \sigma)}$ is on the outside of $\mathcal{D}_{L^2}$. We can choose a new way to do it. Theorem 5.7.2 indicates such a manner. Namely, assume that $T$ is in $\mathcal{D}'_{L^2}$, then $CT(\cdot + i\epsilon) - CT(\cdot - i\epsilon)$ tends to $T$ as $\epsilon \to 0$ in $\mathcal{D}'_{L^2}$. Therefore $CT(\cdot + i\epsilon) - CT(\cdot - i\epsilon)$ converges to $T$ in $S'$. By the continuity of the Fourier transformation in $S'$ we infer that $\widehat{CT}(\cdot + i\epsilon) - \widehat{CT}(\cdot - i\epsilon)$ converges to $\widehat{T}$ in $S'$. Thus, we obtain the following

THEOREM 6.12.1. *If $T$ is in $\mathcal{D}'_{L^2}$ and $CT$ is its Cauchy transform, then*

$$\widehat{T} = \lim_{\epsilon \to 0} (\widehat{CT}(\cdot + i\epsilon) - \widehat{CT}(\cdot - i\epsilon)). \qquad (6.12.1)$$

EXAMPLE 6.12.1. We shall now determine the Fourier transforms of $\frac{1}{(\cdot + i0)^k}$ and $\frac{1}{(\cdot - i0)^k}$. Applying the theory of residues one obtains the following results

$$\mathcal{F}\frac{1}{(\cdot + i\epsilon)^k}(\sigma) = -2\pi i \frac{(i\sigma)^{k-1}}{(k-1)!} e^{\epsilon\sigma} H(-\sigma), \qquad (6.12.2)$$

$$\mathcal{F}\frac{1}{(\cdot - i\epsilon)^k}(\sigma) = 2\pi i \frac{(i\sigma)^{k-1}}{(k-1)!} e^{-\epsilon\sigma} H(\sigma), \qquad (6.12.3)$$

where $H$ is the Heaviside function. Note that

$$-2\pi i \frac{(i \cdot)^{k-1}}{(k-1)!} e^{\epsilon(\cdot)} H(-\cdot) \rightarrow -2\pi i \frac{(i \cdot)^{k-1}}{(k-1)!} H(-\cdot) = \mathcal{F} \frac{1}{(\cdot + i0)^k}$$

in $\mathcal{S}'$ as $\epsilon \rightarrow 0$. Similarly,

$$2\pi i \frac{(i \cdot)^{k-1}}{(k-1)!} e^{-\epsilon(\cdot)} H(\cdot) \rightarrow 2\pi i \frac{(i \cdot)^{k-1}}{(k-1)!} H(\cdot) = \mathcal{F} \frac{1}{(\cdot - i0)^k}$$

as $\epsilon \rightarrow 0$. We are now in a position to determine the Fourier transform of $\frac{1}{(\cdot)^k}$. Since

$$\frac{1}{(\cdot)^k} = \frac{1}{2} \left( \frac{1}{(\cdot + i0)^k} + \frac{1}{(\cdot - i0)^k} \right)$$

therefore

$$\mathcal{F} \frac{1}{(\cdot)^k}(\sigma) = i^k \frac{\pi}{(k-1)!} \sigma^{k-1} \operatorname{sgn} \sigma. \tag{6.12.4}$$

In particular, for $k = 1$ we have

$$\mathcal{F} \frac{1}{(\cdot)}(\sigma) = i\pi \operatorname{sgn} \sigma. \tag{6.12.5}$$

Now, as an immediate consequence of (6.12.2), (6.12.3) and (6.11.1) we obtain the following formula

$$\frac{1}{(\cdot + i0)^\mu} * \frac{1}{(\cdot + i0)^\nu} = -2\pi i \frac{(\mu + \nu - 2)!}{(\mu - 1)!(\nu - 1)!} \frac{1}{(\cdot + i0)^{\mu+\nu-1}}, \tag{6.12.6}$$

$$\frac{1}{(\cdot - i0)^\mu} * \frac{1}{(\cdot - i0)^\nu} = 2\pi i \frac{(\mu + \nu - 2)!}{(\mu - 1)!(\nu - 1)!} \frac{1}{(\cdot - i0)^{\mu+\nu-1}}, \tag{6.12.7}$$

$$\frac{1}{(\cdot - i0)^\mu} * \frac{1}{(\cdot + i0)^\nu} = 0, \tag{6.12.8}$$

where $\mu, \nu \in \mathbb{N}$.

If for some distributions $S$ and $T$ the right side of (6.11.5) is meaningful, then we can define their product by means of (6.11.5).

EXAMPLE 6.12.2. It is easy to check that $\mathcal{F}\{1\}(\sigma) = 2\pi\delta$. By (6.9.1), $\mathcal{F}((\cdot)) = -2\pi i\delta^{(1)}$. We know that $\mathcal{F} \frac{1}{(\cdot)}(\sigma) = i\pi \operatorname{sgn} \sigma$. Note that

$$(i\pi \operatorname{sgn}(\cdot) * (-2\pi i\delta))^{(1)} = (2\pi^2 \operatorname{sgn}(\cdot))^{(1)} = 4\pi^2\delta.$$

Hence we have

$$i\pi \operatorname{sgn}(\cdot) * (-2\pi i\delta)^{(1)} = 2\pi\mathcal{F}\{1\}.$$

Therefore $\frac{1}{(\cdot)}(\cdot) = 1$.

EXAMPLE 6.12.3. According to (6.12.3) we have

$$\left( (\mathcal{F} \frac{1}{(\cdot - i0)^\mu} * \mathcal{F} \frac{1}{(\cdot - i0)^\nu} \right)(\sigma) = 4\pi^2 i^{\mu+\nu} \frac{\sigma^{\mu+\nu-1}}{(\mu + \nu - 1)!}$$

$$= 2\pi \cdot 2\pi i \frac{(i\sigma)^{\mu+\nu-1}}{(\mu + \nu - 1)!} = 2\pi\mathcal{F} \frac{1}{(\cdot + i0)^{\mu+\nu}}(\sigma).$$

From this, by (6.11.5) we get

$$\frac{1}{(\cdot - i0)^\mu} \frac{1}{(\cdot - i0)^\nu} = \frac{1}{(\cdot - i0)^{\mu+\nu}}.$$

Similarly one can show that

$$\frac{1}{(\cdot + i0)^{\mu}} \frac{1}{(\cdot + i0)^{\nu}} = \frac{1}{(\cdot + i0)^{\mu+\nu}}.$$

In particular, for $\mu = \nu = 1$ we have

$$\left(\frac{1}{\cdot - i0}\right)^2 = \frac{1}{(\cdot - i0)^2}.$$

In view of (5.8.14) we get

$$\left(\pi i \delta + \frac{1}{(\cdot)}\right)^2 = -\pi i \delta^{(1)} + \frac{1}{(\cdot)^2}.$$

### 6.13. Hilbert transforms

In this section we define the Hilbert transform of a distribution and show that the Hilbert transformation maps $\mathcal{D}'_{L^2}$ into itself. Moreover, it will be also proven that Sobolev's spaces $W^{m,2}$ are invariant with respect to this transformation.

DEFINITION 6.13.1. Let $T$ be a distribution. Then the Hilbert transform $\mathcal{H}T$ of $T$ is defined as follows

$$\mathcal{H}T = -\frac{1}{\pi}\frac{1}{(\cdot)} * T. \tag{6.13.1}$$

THEOREM 6.13.1. *The Hilbert transformation is one to one mapping of $\mathcal{D}'_{L^2}$ into itself.*

PROOF. In accordance with (6.12.5), $\mathcal{F}\left(-\frac{1}{\pi}\frac{1}{(\cdot)}\right)(\sigma) = -i\,\mathrm{sgn}\,\sigma$. From this, by (6.11.1) we get

$$\mathcal{F}(\mathcal{H}T)(\sigma) = \mathcal{F}\left(-\frac{1}{\pi}\frac{1}{(\cdot)} * T\right)(\sigma) = -i\,\mathrm{sgn}\,\sigma\mathcal{F}T(\sigma).$$

Since $T$ is in $\mathcal{D}'_{L^2}$, therefore by Theorem 6.11.1 we infer that $\mathcal{F}(\mathcal{H}T)$ is in $\mathcal{S}'_0$ and $\mathcal{H}T$ belongs to $\mathcal{D}'_{L^2}$. Note that

$$\mathcal{F}\left(-\frac{1}{\pi}\frac{1}{(\cdot)} * \frac{1}{\pi}\frac{1}{(\cdot)}\right)(\sigma) = 1$$

for each $\sigma \in \mathbb{R}$. Of course, taking into account (6.13.1) we conclude that the following equalities

$$\mathcal{F}\left(-\frac{1}{\pi}\frac{1}{(\cdot)} * \left(\frac{1}{\pi}\frac{1}{(\cdot)} * T\right)\right) = \mathcal{F}\left(\frac{1}{\pi}\frac{1}{(\cdot)} * \left(-\frac{1}{\pi}\frac{1}{(\cdot)} * T\right)\right)$$

$$= \mathcal{F}\left(\left(\frac{1}{\pi}\frac{1}{(\cdot)} * \frac{-1}{\pi}\frac{1}{(\cdot)}\right) * T\right) = \mathcal{F}T \tag{6.13.2}$$

are true for $T \in \mathcal{D}'_{L^2}$. Put by definition $\mathcal{H}^{-1}T := \frac{1}{\pi}\frac{1}{(\cdot)} * T$. Then equalities (6.13.2), because of Theorem 6.8.1 give us the following identities

$$(\mathcal{H} \circ \mathcal{H}^{-1})T = (\mathcal{H}^{-1} \circ \mathcal{H})T = T \tag{6.13.3}$$

for $T$ belonging to $\mathcal{D}'_{L^2}$. This finishes the proof. $\square$

It is easily seen that $f$ is in $W^{m,2}(\mathbb{R})$ (see Section 1.12 and Theorem 6.11.1) iff $(\cdot)^{\alpha}\mathcal{F}f \in L^2(\mathbb{R})$ for $\alpha = 0, 1, 2, \ldots, m$.

Now, we are able to prove the following

THEOREM 6.13.2. *The Sobolev space* $W^{m,2}(\mathbb{R})$, $m = 0, 1, 2 \ldots$ *is invariant with respect to the Hilbert transformation and*

$$\|f\|_{m,2,\mathbb{R}} = \|\mathcal{H}f\|_{m,2,\mathbb{R}} \tag{6.13.4}$$

*for* $f \in W^{m,2}(\mathbb{R})$.

PROOF. Note that

$$D^\alpha(\mathcal{H}f) = -\frac{1}{\pi} \frac{1}{(\cdot)} * D^\alpha f.$$

Hence, we have

$$\mathcal{F}(D^\alpha(\mathcal{H}f)) = -i \operatorname{sgn} \sigma \mathcal{F}(D^\alpha f)(\sigma) = -i \operatorname{sgn} \sigma(-i\sigma)^\alpha \mathcal{F}f(\sigma).$$

We see at once that $\mathcal{F}(D^\alpha(\mathcal{H}f)) \in L^2(\mathbb{R})$ if $\alpha = 0, 1, \ldots, m$. By Theorem 6.6.2 we infer that $\mathcal{H}f \in W^{m,2}(\mathbb{R})$. Equality (6.13.4) is evident. $\qquad\square$

## 6.14. Carleman transforms

T. Carleman defined in [4] a transform of a slowly increasing function as follows

DEFINITION 6.14.1. Let $f$ be in $\mathcal{S}_0'(\mathbb{R})$. Set

$$\widehat{\mathcal{F}}f(z) = \begin{cases} \int_0^\infty f(t)e^{itz}\,dt & \text{if } \Im z > 0, \\ -\int_{-\infty}^0 f(t)e^{itz}\,dt & \text{if } \Im z < 0. \end{cases}$$

Similarly, we put

$$\widehat{\mathcal{F}}^{-1}f(z) = \begin{cases} \frac{1}{2\pi}\int_{-\infty}^0 f(t)e^{-itz}\,dt & \text{if } \Im z > 0, \\ -\frac{1}{2\pi}\int_0^\infty f(t)e^{-itz}\,dt & \text{if } \Im z < 0. \end{cases}$$

We shall now prove the following

THEOREM 6.14.1. *If $f$ is in $\mathcal{S}_0'$, then*

$$\widehat{\mathcal{F}}f(z) = (C \circ \mathcal{F})f(z) \tag{6.14.1}$$

*for $z \in \mathbb{C} - \mathbb{R}$, where the symbols $\mathcal{F}$ and $C$ denote the Fourier transformation and the Cauchy transformation, respectively.*

PROOF. First we prove a reduced form of this theorem. Namely, we assume that $f$ is in $L^2(\mathbb{R})$. Determine $(C \circ \mathcal{F})f(z)$ for $z \in \mathbb{C} - \mathbb{R}$. According to the definitions of $C$ and $\mathcal{F}$ we have

$$(C \circ \mathcal{F})f(z) = \frac{1}{2\pi i} \int_{\mathbb{R}} \mathcal{F}f(t)\frac{1}{t - z}\,dt.$$

Applying Theorem 6.6.3 we get

$$(C \circ \mathcal{F})f(z) = \frac{1}{2\pi i} \int_{\mathbb{R}} f(t)\mathcal{F}\frac{1}{(\cdot - z)}(t)\,dt.$$

It is known (see Example 6.6.2) that

$$\frac{1}{2\pi i}\mathcal{F}\frac{1}{(\cdot - z)}(t) = \begin{cases} H(t)e^{itz} & \text{if } \Im z > 0, \\ -H(-t)e^{itz} & \text{if } \Im z < 0. \end{cases}$$

From this we obtain

$$(C \circ \mathcal{F})f(z) = \begin{cases} \int_0^\infty f(t)e^{itz}\,dt & \text{if } \Im z > 0, \\ -\int_{-\infty}^0 f(t)e^{itz}\,dt & \text{if } \Im z < 0. \end{cases}$$

Finally we have (6.14.1) for $f \in L^2(\mathbb{R})$.

We are able to show that the theorem is true in general. It remains to prove that (6.14.1) holds for $f(x) = (ix)^k f_0(x)$, where $f_0 \in L^2(\mathbb{R})$ and $k = 0, 1, 2, \ldots$ In fact, we have

$$(C \circ \mathcal{F})((i\cdot)^k f_0(\cdot))(z) = C((\mathcal{F}f_0)^{(k)})(z) = ((C \circ \mathcal{F})f_0)^{(k)}(z) = \frac{d^k}{dz^k}\widehat{\mathcal{F}}f_0(z)$$

$$= \begin{cases} \int_0^\infty e^{itz}(it)^k f_0(t)\,dt & \text{if } \Im z > 0, \\ -\int_{-\infty}^0 e^{itz}(it)^k f_0(t)\,dt & \text{if } \Im z < 0. \end{cases}$$

Taking into account Definition 6.11.1 we get (6.14.1) for $f \in \mathcal{S}_0'$. In the same way one can prove the following equality

$$\widehat{\mathcal{F}}^{-1}f(z) = (C \circ \mathcal{F}^{-1})f(z) \tag{6.14.2}$$

for $f \in \mathcal{S}_0'$.                                                                           $\square$

As a corollary from Theorem 5.7.1, Theorem 5.7.2 and Theorem 6.11.1 we obtain

THEOREM 6.14.2. *If $f \in \mathcal{S}_0'$, then $\widehat{\mathcal{F}}f$ is a holomorphic function in $\mathbb{C} - \mathbb{R}$ and*

$$\lim_{\epsilon \to 0} \int_{\mathbb{R}} (\widehat{\mathcal{F}}f(x + i\epsilon) - \widehat{\mathcal{F}}f(x - i\epsilon))\varphi(x)\,dx = \mathcal{F}f(\varphi) \tag{6.14.3}$$

*for $\varphi \in \mathcal{D}_{L^2}$.*

Formula (6.14.3) may be used to determine the Fourier transforms of slowly increasing functions.

EXAMPLE 6.14.1. Of course, the Heaviside function $H$ is slowly increasing. It is easy to see that

$$\widehat{\mathcal{F}}H(z) = \begin{cases} -\frac{1}{iz} & \text{if } \Im z > 0, \\ 0 & \text{if } \Im z < 0. \end{cases}$$

Therefore

$$\mathcal{F}H(\varphi) = -\lim_{\epsilon \to 0} \frac{1}{i} \int_{\mathbb{R}} \frac{1}{x + i\epsilon}\varphi(x)\,dx = i\frac{1}{(\cdot + i0)}(\varphi).$$

REMARK 6.14.1. If $f \in L^2$ then

$$\lim_{\epsilon \to 0}(\widehat{\mathcal{F}}f(\cdot + i\epsilon) - \widehat{\mathcal{F}}f(\cdot - i\epsilon)) = \mathcal{F}f$$

in $L^2$.

## 6.15. The Paleya–Wiener type theorems

We shall now recall the classical Paleya–Wiener theorem ([23]).

THEOREM 6.15.1. *Let $f \in L^2(0, \infty)$ and*

$$F(z) = \int_0^\infty e^{izt} f(t)\, dt, \tag{6.15.1}$$

*then $F$ is a holomorphic function in $\Pi^+ = \{z : \Im z > 0\}$ and*

$$\sup_{0 < y < \infty} \int_{\mathbb{R}} |F(x + iy)|^2\, dx = C < \infty. \tag{6.15.2}$$

*Conversely, if $F$ is a holomorphic function in $\Pi^+$ and (6.15.2) holds then there exists $f \in L^2(0, \infty)$ such that (6.15.1) is true for $z \in \Pi^+$. Similarly, if $f \in L^2(-\infty, 0)$ and*

$$F(z) = -\int_{-\infty}^0 e^{izt} f(t)\, dt, \tag{6.15.3}$$

*then $F$ is a holomorphic function in $\Pi^- = \{z : \Im z < 0\}$ and*

$$\sup_{-\infty < y < 0} \int_{\mathbb{R}} |F(x + iy)|^2\, dx = C < \infty. \tag{6.15.4}$$

*Conversely, if $F$ is a holomorphic function in $\Pi^-$ and (6.15.4) holds then there exists $f \in L^2(-\infty, 0)$ such that (6.15.3) is true for $z \in \Pi^-$.*

Denote by $H(\Pi^- \cup \Pi^+)$ the vector space of all holomorphic functions in $\Pi^- \cup \Pi^+$. As a corollary we obtain the following

THEOREM 6.15.2. *$F \in H(\Pi^- \cup \Pi^+)$ and*

$$\sup_{y \neq 0} \int_{\mathbb{R}} |F(x + iy)|^2\, dx = C < \infty \tag{6.15.5}$$

*iff there exists a function $f$ in $L^2(\mathbb{R})$ such that*

$$F(z) = \widehat{\mathcal{F}} f(z)$$

*for $z \in \Pi^- \cup \Pi^+$, where $\widehat{\mathcal{F}} f$ is the Carleman transform of $f$.*

We are able to prove a similar theorem for the Cauchy transform.

THEOREM 6.15.3. *$F$ is in $H(\Pi^- \cup \Pi^+)$ and (6.15.5) holds, iff there exists a function $f \in L^2(\mathbb{R})$ such that*

$$F(z) = Cf(z)$$

*for $z \in (\Pi^- \cup \Pi^+)$, where $Cf$ is the Cauchy transform of $f$.*

PROOF. By Theorem 6.15.2 there exists a function $g$ in $L^2(\mathbb{R})$ such that $F(z) = \widehat{\mathcal{F}} g(z)$. Taking into account (6.14.1) we get $\widehat{\mathcal{F}} g(z) = (C \circ \mathcal{F}) g(z)$. Put $f = \mathcal{F} g$, then we have $F(z) = Cf(z)$. This finishes the proof. $\qquad\square$

We are now in a position to give a generalization of this theorem.

THEOREM 6.15.4. *A function $F \in H(\Pi^- \cup \Pi^+)$ is the Cauchy transform of some distribution $\Lambda$ from $\mathcal{D}'_{L^2}$ iff*

$$F(z) = \sum_{\nu=0}^{m} \frac{d^\nu}{dz^\nu} F_\nu(z), \qquad (6.15.6)$$

*where $F_\nu \in H(\Pi^- \cup \Pi^+)$ and fulfils (6.15.5).*

PROOF OF NECESSITY. Let $\Lambda \in \mathcal{D}'_{L^2}$, then we have

$$C\Lambda(z) = \frac{1}{2\pi i} \Lambda\left(\frac{1}{(\cdot - z)}\right).$$

In view of Theorem 5.4.1 we obtain the following equality

$$C\Lambda(z) = \frac{1}{2\pi i} \sum_{\nu=0}^{m} (-1)^\nu \int_{\mathbb{R}} f_\nu(t) \frac{\partial^\nu}{\partial t^\nu} \frac{1}{t-z} dt,$$

where $f_\nu \in L^2(\mathbb{R})$. This means that $C\Lambda(z) = \sum_{\nu=0}^{m} C D^\nu f_\nu(z)$, where $D^\nu f_\nu$ denotes $\nu$-distributional derivative of $f_\nu$. In accordance with Theorem 5.7.1 we have

$$C\Lambda(z) = \sum_{\nu=0}^{m} \frac{d^\nu}{dz^\nu} C f_\nu(z).$$

Put $F_\nu(z) := C f_\nu(z)$. By virtue of Theorem 6.15.3, $F_\nu$ fulfils (6.15.5). $\qquad \square$

PROOF OF SUFFICIENCY. Suppose that $F(z) = \sum_{\nu=0}^{m} \frac{d^\nu}{dz^\nu} F_\nu(z)$, where the function $F_\nu$ fulfils (6.15.5), $\nu = 1, \ldots, m$. In view of Theorem 6.15.3 we infer that there exists a function $f_\nu \in L^2(\mathbb{R})$ such that $F_\nu = C f_\nu$. Therefore

$$F(z) = \sum_{\nu=0}^{m} \frac{d^\nu}{dz^\nu} C f_\nu(z) = \sum_{\nu=0}^{m} C(D^\nu f_\nu)(z).$$

This means that $F = C\Lambda$, where $\Lambda = \sum_{\nu=0}^{m} D^\nu f_\nu$, $f_\nu \in L^2(\mathbb{R})$. This implies that $\Lambda$ is in $\mathcal{D}'_{L^2}$. $\qquad \square$

We are able to form a similar theorem for the Carleman transformation.

THEOREM 6.15.5. *A function $F$ in $H(\Pi^- \cup \Pi^+)$ is the Carleman transform of a slowly increasing function $f$ iff $F$ may be written in the form (6.15.6), where the function $F_\nu$ fulfils (6.15.5).*

PROOF. The theorem is an immediate consequence of the previous theorem and (6.14.1). $\qquad \square$

EXAMPLE 6.15.1. Let $\chi_{[-1,1]}$ be a characteristic function of the interval $[-1,1]$. An easy computation shows that

$$\mathcal{F}\left(\frac{1}{2}\chi_{[-1,1]}\right)(x) = \frac{\sin x}{x}$$

(see Example 6.3.1) and

$$\hat{\mathcal{F}}\left(\frac{1}{2}\chi_{[-1,1]}\right)(z) = \begin{cases} \frac{e^{iz}-1}{2iz} & \text{if } \Im z > 0, \\ \frac{e^{-iz}-1}{2iz} & \text{if } \Im z < 0. \end{cases}$$

Hence we have

$$\frac{1}{2\pi i} \int\limits_{\mathbb{R}} \frac{1}{x - z} \frac{\sin x}{x}\, dx = \begin{cases} \frac{e^{iz}-1}{2iz} & \text{if } \Im z > 0, \\ \frac{e^{-iz}-1}{2iz} & \text{if } \Im z < 0. \end{cases}$$

EXAMPLE 6.15.2. It is easy to verify that the Carleman transform of the Heaviside function is

$$\widehat{\mathcal{F}}H(z) = \begin{cases} \frac{i}{z} & \text{if } \Im z > 0, \\ 0 & \text{if } \Im z < 0. \end{cases}$$

In accordance with Theorem 6.14.2

$$\lim_{y \to 0^+} i \int\limits_{\mathbb{R}} \frac{1}{x + iy} \varphi(x)\, dx = \frac{i}{(\cdot + i0)}(\varphi)$$

if $\varphi \in \mathcal{D}_{L^2}$. Therefore

$$C\frac{1}{(\cdot + i0)}(z) = \begin{cases} \frac{i}{z} & \text{if } \Im z > 0, \\ 0 & \text{if } \Im z < 0. \end{cases}$$

## 6.16. The Cauchy semigroup

Let $g_t(x) := \frac{1}{\pi} \frac{t}{t^2 + x^2}$, $x \in \mathbb{R}$ and $t > 0$. Assume that $\varphi \in \mathcal{S}(\mathbb{R})$ and put $u_t(\varphi) := g_t * \varphi$. By Young's inequality we have

$$\|u_t(\varphi)\|_{L^2} \leq \|g_t\|_{L^1} \|\varphi\|_{L^2} = \|\varphi\|_{L^2}.$$

This implies that $u_t$ is a contraction operator on $L^2(\mathbb{R})$ for $t > 0$. We know that $\mathcal{F}g_t(\sigma) = e^{-t|\sigma|}$. Using the classical Borel formula one can prove that

$$g_{t_1} * g_{t_2} = g_{t_1 + t_2}$$

(see (6.4.7)). From this we get

$$u_{t_1 + t_2}(\varphi) = (g_{t_1} * g_{t_2}) * \varphi = g_{t_1} * (g_{t_2} * \varphi) = u_{t_1}(u_{t_2}(\varphi))$$

for $\varphi \in \mathcal{S}(\mathbb{R})$. This means that $u_t(\cdot)$ is a semigroup of the contractions $u_t$, $t > 0$ on $L^2(\mathbb{R})$. This semigroup is called the Cauchy semigroup. Our purpose is to find the infinitesimal operator of this semigroup. Let us notice that

$$\mathcal{F}(t^{-1}(u_t(\varphi) - \varphi))(\sigma) = t^{-1}(e^{-t|\sigma|} - 1)\mathcal{F}\varphi(\sigma).$$

It is easy to see that $|t^{-1}(e^{-t|\sigma|} - 1)| \leq |\sigma|$. Therefore

$$|t^{-1}(e^{-t|\sigma|} - 1)\mathcal{F}\varphi(\sigma)| + |\sigma|\,|\mathcal{F}\varphi(\sigma)| \leq 2|\sigma|\,|\mathcal{F}\varphi(\sigma)|.$$

Of course, $(\cdot)\mathcal{F}\varphi$ is in $L^2(\mathbb{R})$ if $\varphi$ belongs to $\mathcal{S}(\mathbb{R})$. On the other hand $t^{-1}(e^{-t|\sigma|} - 1)$ tends pointwise to $-|\sigma|$ as $t \to 0^+$. This implies that

$$\mathcal{F}(t^{-1}(u_t(\varphi) - \varphi)) \to -|(\cdot)|\mathcal{F}\varphi$$

as $t \to 0$ in $L^2(\mathbb{R})$. Using the Plancherel theorem we infer that $t^{-1}(u_t(\varphi) - \varphi)$ converges to some $\psi$ in $L^2(\mathbb{R})$. Obviously, $\mathcal{F}\psi(\sigma) = -|\sigma|\mathcal{F}\varphi(\sigma)$. In view of Theorem 6.10.2 we have

$$\mathcal{F}\left(\frac{1}{\pi} \frac{1}{(\cdot)^2} * \varphi\right)(\sigma) = -|\sigma|\mathcal{F}\varphi(\sigma).$$

Of course, $\frac{1}{\pi} \frac{1}{(\cdot)^2} * \varphi = \psi$ in $L^2(\mathbb{R})$. From this it follows that the operator

$$A(\varphi) := \frac{1}{\pi} \frac{1}{(\cdot)^2} * \varphi$$

is the infinitesimal operator of the Cauchy semigroup if we take $\mathcal{S}(\mathbb{R})$ as the domain of $u_t$, $t > 0$.

The principal goal of our consideration is to show that the Cauchy semigroup may be reproduced by means of the exponential function $\exp A(\cdot)$ on the dense subspace $\overset{\circ}{\mathcal{S}}$ of $L^2(\mathbb{R})$. We shall now define such subspace of $L^2(\mathbb{R})$. In order to do it we use the Hermite functions $h_\nu$ (see Section 6.7). Let $\overset{\circ}{\mathcal{S}}$ denote the vector space spanned by the set $\{h_\nu : \nu = 1, 2, \ldots\}$. Since Hermite functions $h_\nu$ constitute a complete orthonormal system in $L^2(\mathbb{R})$, therefore the closure of $\overset{\circ}{\mathcal{S}}$ in $L^2(\mathbb{R})$ is the whole space $L^2(\mathbb{R})$. We know that

$$\mathcal{F}h_k = i^k \sqrt{2\pi} h_k.$$

For simplicity put $\chi := \frac{1}{\pi} \frac{1}{(\cdot)^2}$. Let $\chi^\nu = \underbrace{\chi * \cdots * i}_{\nu \text{ times}}$. In accordance with Theorem 6.10.2 and (6.12.4) we have

$$\mathcal{F}(\chi^\nu * h_k) = (-1)^\nu |\sigma|^\nu i^k \sqrt{2\pi} h_k(\sigma).$$

From this and the Plancherel theorem it follows that $\chi^\nu * h_k$ is in $\mathcal{D}_{L^2}$. We shall now show that the series

$$h_k + \frac{t}{1!} \chi * h_k + \frac{t^2}{2!} \chi^2 * h_k + \cdots + \frac{t^n}{n!} \chi^n * h_k + \cdots \qquad (6.16.1)$$

converges to $u_t(h_k)$ in the sense of $L^2(\mathbb{R})$ for $k = 0, 1, \ldots$ First let us find the Fourier transforms of the partial sums

$$S_{nt}^k = h_k + \frac{t}{1!} \chi * h_k + \cdots + \frac{t^n}{n!} \chi^n * h_k$$

of the above series. By virtue of the Borel formula we obtain

$$\mathcal{F}S_{nt}^k = i^k \sqrt{2\pi} h_k(\sigma) \left(1 - \frac{t}{1!}|\sigma| + \cdots + (-1)^n \frac{t^n}{n!}|\sigma|^n\right).$$

Note that

$$|\mathcal{F}S_{nt}^k(\sigma) - i^k \sqrt{2\pi} h_k(\sigma) e^{-t|\sigma|}|^2 \leq 2\pi h_k^2(|\sigma|) \left|\sum_{\nu=n+1}^{\infty} (-1)^\nu \frac{t^\nu}{\nu!}|\sigma|^\nu\right|^2$$

$$\leq C_k e^{-|\sigma|^2} e^{2t(|\sigma|+\epsilon)} \in L^1(\mathbb{R}),$$

for $t > 0$ and some $\epsilon > 0$. Obviously, $\mathcal{F}S_{nt}^k(\sigma)$ converges to $2\pi i^k h_k(\sigma) e^{-t|\sigma|}$ for $\sigma \in \mathbb{R}$. Therefore, by the Lebesgue dominated convergence theorem we conclude that $S_{nt}^k$ converges to $\sqrt{2\pi} i^k h_k e^{-t|(\cdot)|}$ as $n \to \infty$ in the sense of $L^2$. Because of the Plancherel theorem the series (6.16.1) converges to $u_t(h_k)$ in $L^2$. Putting for simplicity

$$e^{At} h_k := \lim_{n \to \infty} \sum_{\nu=0}^{n} \frac{\chi^\nu t^\nu}{\nu!} * h_k,$$

we finally arrive at the following

THEOREM 6.16.1. *If $\varphi \in \overset{\circ}{\mathcal{S}}$, then $u_t(\varphi) = e^{At}\varphi$ in the sense of $L^2$.*

REMARK 6.16.1. If we take $W^{1,2}(\mathbb{R})$ as the domain of the operator $A$, then it will be a closed operator. We only need to verify that if $\varphi_\nu$ are in $W^{1,2}(\mathbb{R})$, $\varphi_\nu \to \varphi$ in $L^2$ and $A\varphi_\nu \to \psi$ in $L^2$, then $\varphi$ belongs to $W^{1,2}(\mathbb{R})$ and $A\varphi = \psi$. By Theorem 6.11.2 we have

$$\widehat{\frac{1}{(\cdot)^2} * \varphi_\nu} = -|(\cdot)|\widehat{\varphi_\nu}.$$

In accordance with Theorem 6.6.2, $\widehat{\varphi_\nu} \to \widehat{\varphi}$ in $L^2$ and $-|(\cdot)|\widehat{\varphi_\nu} \to \widehat{\psi}$ in $L^2$. One can choose a subsequence $(\widehat{\varphi_{\nu_k}})$ of the sequence $(\widehat{\varphi_\nu})$ so that $\widehat{\varphi_{\nu_k}} \to \widehat{\varphi}$, $-|(\cdot)|\widehat{\varphi_{\nu_k}} \to \widehat{\psi}$ and $-|(\cdot)|\widehat{\varphi_{\nu_k}} \to -|(\cdot)|\widehat{\varphi}$ as $\nu \to \infty$ almost everywhere on $\mathbb{R}$. Therefore $-|(\cdot)|\widehat{\varphi} = \widehat{\psi}$ in $L^2(\mathbb{R})$. This implies that $\varphi$ is in $W^{1,2}(\mathbb{R})$ and $A\varphi = \psi$.

## 6.17. The Cauchy problem for the heat equation

A formal solution of the Cauchy problem

$$\frac{\partial}{\partial t}u(x,t) = \Delta_x u(x,t), \qquad (6.17.1)$$

$$u(x,0) = f(x), \qquad (6.17.2)$$

where $x \in \mathbb{R}^n$, $t \geq 0$ and $\Delta_x = \frac{\partial^2}{\partial x_1^2} + \cdots + \frac{\partial^2}{\partial x_n^2}$, can be constructed by quadratures. The solution of equation (6.17.1), which satisfies (6.17.2) is usually written as follows

$$u(x,t) = (4\pi t)^{-\frac{n}{2}} \int_{\mathbb{R}^n} \exp\frac{-|x-y|^2}{4t} f(y)\, dy. \qquad (6.17.3)$$

Then $f(x) = \lim_{t\to 0+} u(x,t)$. The function

$$E(x,t) = \begin{cases} (4\pi t)^{-\frac{n}{2}} \exp\left(\frac{-|x|^2}{4t}\right) & \text{if } t > 0, \\ 0 & \text{if } t \leq 0 \end{cases}$$

is a distributional solution of the differential equation

$$\left(\Delta_x - \frac{\partial}{\partial t}\right) E = -\delta, \qquad (6.17.4)$$

where $\delta$ denotes the Dirac distribution (compare with Section 4.3). Equality (6.17.3) can be written in the convolution form

$$u(x,t) = (E(\cdot,t) * f)(x). \qquad (6.17.5)$$

We shall now show that this formula holds when we substitute an arbitrary tempered distribution $\Lambda$ in the place of $f$. In this case $u(\cdot,\cdot)$ is also a smooth function in $\Omega = \{(x,t) : x \in \mathbb{R}^n, \ t > 0\}$. A question, how the compliance of (6.17.2) can be understood, when a tempered distribution $\Lambda$ is taken in place of the function $f$, naturally arises. In this sequel we explain the above question.

We start with the following

LEMMA 6.17.1. *Let $f$ and $g$ be in $\mathcal{S}(\mathbb{R}^n)$. Put $h(x,\sigma) := f(x-\sigma)g(\sigma)$, then $h \in \mathcal{S}(\mathbb{R}^n \times \mathbb{R}^n)$.*

PROOF. Note that

$$\frac{\partial^{|\alpha|}}{\partial\sigma^\alpha}h(x,\sigma) = \sum_{0\leq\nu\leq\alpha} \binom{\alpha}{\nu} \frac{\partial^{|\nu|}}{\partial\sigma^\nu}f(x-\sigma)\frac{\partial^{|\alpha-\nu|}}{\partial\sigma^{\alpha-\nu}}g(\sigma), \quad \alpha \in \mathbb{N}^n.$$

Hence we obtain

$$\sigma^{\beta_1} x^{\beta_2} \frac{\partial^{|\alpha_1+\alpha_2|}}{\partial\sigma^{\alpha_1}\partial x^{\alpha_2}} h(x,\sigma) = \sum_{0\leq\nu\leq\alpha_1} \binom{\alpha_1}{\nu} x^{\beta_2} \frac{\partial^{|\alpha_2+\nu|}}{\partial x^{\alpha_2}\partial\sigma^{\nu}} f(x-\sigma)\sigma^{\beta_1} \frac{\partial^{|\alpha_1-\nu|}}{\partial\sigma^{\alpha_1-\nu}} g(\sigma).$$

By means of the Taylor formula we get $x^{\beta_2} = w(x-\sigma,\sigma)$, where $w(\cdot,\cdot)$ is a polynomial. Therefore

$$x^{\beta_2} \frac{\partial^{|\alpha_2+\nu|}}{\partial x^{\alpha_2}\partial\sigma^{\nu}} f(x-\sigma) = w(x-\sigma,\sigma)\frac{\partial^{|\alpha_2+\nu|}}{\partial x^{\alpha_2}\partial\sigma^{\nu}} f(x-\sigma).$$

It is easily seen that

$$\frac{\partial^{|\alpha_2+\nu|}}{\partial x^{\alpha_2}\partial\sigma^{\nu}} f(x-\sigma) = (-1)^{|\nu|}\frac{\partial^{|\alpha_2+\nu|}}{\partial y^{\alpha_2+\nu}} f(y),$$

where $y = x - \sigma$. The expression

$$w(y,\sigma)\sigma^{\beta_1}(-1)^{|\nu|}\frac{\partial^{|\alpha_2+\nu|}}{\partial y^{\alpha_2+\nu}} f(y)\frac{\partial^{|\alpha_1-\nu|}}{\partial\sigma^{\alpha_1-\nu}} g(\sigma)$$

is bounded in $\mathbb{R}^n \times \mathbb{R}^n$ by the assumption. This implies that the function

$$\sigma^{\beta_1} x^{\beta_2} \frac{\partial^{|\alpha_1+\alpha_2|}}{\partial\sigma^{\alpha_1}\partial x^{\alpha_2}} h(x,\sigma)$$

is bounded in $\mathbb{R}^n \times \mathbb{R}^n$. Thus, the proof is finished. $\qquad\square$

We shall now show that the function

$$u(x,\sigma) := \Lambda_\sigma(E(x-\sigma,t)) \tag{6.17.6}$$

satisfies equation (6.17.1) in $\Omega$. Indeed, we have to show that

$$\int\limits_{\mathbb{R}^+}\int\limits_{\mathbb{R}^n} \Lambda_\sigma(E(x-\sigma,t))\left(\Delta_x + \frac{\partial}{\partial t}\right)\varphi(x,t)\,dt\,dx = 0 \tag{6.17.7}$$

for $\varphi \in \mathcal{D}(\Omega)$. In view of Lemma 6.17.1 we infer that

$$E(x-\sigma,t)\left(\Delta_x + \frac{\partial}{\partial t}\right)\varphi(x,t) \in \mathcal{S}(\mathbb{R}^n \times \mathbb{R}^n \times \mathbb{R}).$$

Moreover, the characteristic function $\Gamma$ of the set $\mathbb{R}^+ \times \mathbb{R}^n$ belongs to $\mathcal{S}'(\mathbb{R} \times \mathbb{R}^n)$. Equality (6.17.7) can be written in the form of the tensor product of the distributions $\Gamma$ and $\Lambda$ as follows

$$(\Gamma_{(t,x)} \otimes \Lambda_\sigma)\left(E(x-\sigma,t)\left(\Delta_x + \frac{\partial}{\partial t}\right)\varphi(x,t)\right) = 0 \tag{6.17.8}$$

(see Example 6.2.1). By commutativity of the tensor product of tempered distributions we have

$$(\Gamma_{(t,x)} \otimes \Lambda_\sigma)\left(E(x-\sigma,t)\left(\Delta_x + \frac{\partial}{\partial t}\right)\varphi(x,t)\right)$$

$$= \Lambda_\sigma\left(\int\limits_{\mathbb{R}^+}\int\limits_{\mathbb{R}^n} E(x-\sigma,t)\left(\Delta_x + \frac{\partial}{\partial t}\right)\varphi(x,t)\,dt\,dx\right).$$

The integral

$$\int\limits_{\mathbb{R}^+}\int\limits_{\mathbb{R}^n} E(x-\sigma,t)\left(\Delta_x + \frac{\partial}{\partial t}\right)\varphi(x,t)\,dt\,dx$$

can be written in the following form

$$\int\int_{\mathbb{R}^+ \,\mathbb{R}^n} \left( \Delta_x - \frac{\partial}{\partial t} \right) E(x - \sigma, t) \varphi(x, t) \, dt \, dx.$$

By (6.17.4) we have

$$\int\int_{\mathbb{R}^+ \,\mathbb{R}^n} \left( \Delta_x - \frac{\partial}{\partial t} \right) E(x - \sigma, t) \varphi(x, t) \, dt \, dx = 0$$

for $\sigma \in \mathbb{R}^n$. According to Theorem 4.6.4, $u(\cdot, \cdot)$ is smooth in $\Omega$.

We have proved the fact which can be stated as a separate

THEOREM 6.17.1. *A function* $u(x, t) = \Lambda_\sigma(E(x - \sigma, t))$, $\Lambda \in \mathcal{S}'(\mathbb{R}^n)$ *is a solution of* (6.17.1) *in* $\Omega$.

The sense of satisfying (6.17.2) by this solution will be explained by the following

THEOREM 6.17.2. $u(\cdot, t)$ *converges to* $\Lambda$ *in* $\mathcal{S}'(\mathbb{R}^n)$ *as* $t \to 0^+$.

PROOF. Since $\Lambda$ is in $\mathcal{S}'(\mathbb{R}^n)$, therefore

$$|u(x, t)| \leq |\Lambda_\sigma(E(x - \sigma, t))|$$

$$\leq M_{\Lambda t} \sum_{|\alpha| \leq m_1} \sum_{|\beta| \leq m_2} \sup_{\sigma \in \mathbb{R}^n} \left| \sigma^\beta \frac{\partial^{|\alpha|}}{\partial \sigma^\alpha} E(x - \sigma, t) \right| \leq K_{\Lambda t}.$$

We shall now show that

$$\int_{\mathbb{R}^n} u(x, t) \varphi(x) \, dx \to \Lambda(\varphi) \tag{6.17.9}$$

as $t \to 0^+$ for $\varphi \in \mathcal{S}(\mathbb{R}^n)$. In accordance with Lemma 6.17.1, $E(x - \sigma, t) \varphi(x)$ is in $\mathcal{S}(\mathbb{R}^n \times \mathbb{R}^n)$ for $t > 0$. Commutativity of the tensor product of tempered distributions implies the following equality

$$\int_{\mathbb{R}^n} \Lambda_\sigma(E(x - \sigma, t)) \varphi(x) \, dx = \Lambda_\sigma \left( \int_{\mathbb{R}^n} (E(x - \sigma, t) \varphi(x) \, dx) \right)$$

for $\varphi \in \mathcal{S}(\mathbb{R}^n)$. We see at once that

$$\int_{\mathbb{R}^n} E(x - \sigma, t) \varphi(x) \, dx = \int_{\mathbb{R}^n} E(\sigma - x, t) \varphi(x) \, dx = (E(\cdot, t) * \varphi)(\sigma).$$

We have to show that the expression $E(\cdot, t) * \varphi$ converges to $\varphi$ as $t \to 0^+$ in $\mathcal{S}(\mathbb{R}^n)$ for each $\varphi \in \mathcal{S}(\mathbb{R}^n)$. Equivalently,

$$x^\beta \frac{\partial^{|\alpha|}}{\partial x^\alpha} (E(\cdot, t) * \varphi)(x) \to x^\beta \frac{\partial^{|\alpha|}}{\partial x^\alpha} \varphi(x) \tag{6.17.10}$$

uniformly as $t \to 0^+$ in $\mathbb{R}^n$ for each $\alpha, \beta \in \mathbb{N}^n$. It is easy to check that

$$(\cdot)^\beta \partial^\alpha (E(\cdot, t) * \varphi) = (\cdot)^\beta (E(\cdot, t) * \partial^\alpha \varphi)$$

$$= E(\cdot, t) * ((\cdot)^\beta \partial^\alpha \varphi) + \sum_{0 < \nu \leq \beta} \binom{\beta}{\nu} ((\cdot)^\nu E(\cdot, t)) * ((\cdot)^{\beta - \nu} \partial^\alpha \varphi).$$

Hence, it follows that

$$(\cdot)^\beta \partial^\alpha \widehat{(E(\cdot,t) * \varphi)}(\sigma) = (-i)^{|\beta|} \exp(-t|\sigma|^2) \frac{\partial^{|\beta|}}{\partial \sigma^\beta} \widehat{\partial^\alpha \varphi}(\sigma)$$

$$+ (-i)^{|\beta|} \sum_{0 < \nu \le \beta} \binom{\beta}{\nu} \frac{\partial^{|\nu|}}{\partial \sigma^\nu} \exp(-t|\sigma|^2) \frac{\partial^{|\beta-\nu|}}{\partial \sigma^{\beta-\nu}} \widehat{\partial^\alpha \varphi}(\sigma).$$

One can observe that

$$\partial^\nu \exp(-t|\cdot|^2) \partial^{\beta-\nu} \widehat{\partial^\alpha \varphi} \to 0$$

as $t \to 0^+$ in $L^1(\mathbb{R}^n)$, when $\nu > 0$. Of course, Theorem 6.3.1 is also true if we put $\mathcal{F}^{-1}$ in the place of $\mathcal{F}$. Now, it is clear that (6.17.10) holds. □

Assuming that $\Lambda$ belongs to a small subspace of $\mathcal{S}'(\mathbb{R}^n)$ we can obtain a more delicate information about the solution $u$ near the boundary of $\Omega$.

THEOREM 6.17.3. *If $\Lambda = f \in L^p(\mathbb{R}^n)$, $1 \le p < \infty$ and $u(t,x) = \Lambda(E(x - \cdot, t))$, then $u(\cdot, t)$ tends to $\Lambda$ as $t \to 0^+$ in $L^p$.*

PROOF. Note that $E(x,t) = t^{-\frac{n}{2}} E\left(\frac{x}{\sqrt{t}}, 1\right)$ and $\int_{\mathbb{R}^n} E(x,1)\,dx = 1$. Since

$$\Lambda(E(x - \cdot, \sqrt{t})) = \left(t^{-\frac{n}{2}} E\left(\frac{\cdot}{\sqrt{t}}, 1\right) * f\right)(x), \qquad (6.17.11)$$

therefore it is a consequence of Lemma 1.2.1. □

Since $u(\cdot, \cdot)$ is smooth in $\Omega$, therefore $u(\cdot, t)$ does not converge to $f$ as $t \to 0+$ in $L^\infty(\mathbb{R}^n)$, if we put $f = H$ in (6.17.11), where $H$ denotes the Heaviside function. It means that the above theorem is not true, when $p = \infty$.

THEOREM 6.17.4. *If $\Lambda \in \mathcal{D}'_{L^p}$, $1 \le p < \infty$ and $u(x,t) = \Lambda(E(x - \cdot, t))$, then $u(\cdot, t)$ converges to $\Lambda$ in $\mathcal{D}'_{L^p}$ as $t \to 0^+$.*

PROOF. This theorem is an immediate consequence of Theorem 5.3.2 and Lemma 5.7.1. □

CHAPTER 7

# ORTHOGONAL EXPANSIONS OF DISTRIBUTIONS

### 7.1. The Poisson summation formula

Let $T = \sum_{\nu \in \mathbb{Z}} \tau_\nu \delta$. First, we shall show that $T$ is a tempered distribution. Note that $T(\varphi) = \sum_{\nu \in \mathbb{Z}} \varphi(\nu)$. Of course, the series $\sum_{\nu \in \mathbb{Z}} \varphi(\nu)$ is convergent for each $\varphi \in \mathcal{S}$. Since the space $\mathcal{S}'$ is complete (see Section 6.1 and Appendix), therefore $T$ is in $\mathcal{S}'$. Denote by $\widehat{T}$ the Fourier transform of $T$. We are able to prove the following

THEOREM 7.1.1.

$$\widehat{\sum_{\nu \in \mathbb{Z}} \tau_\nu \delta} = 2\pi \sum_{\nu \in \mathbb{Z}} \tau_{2\pi\nu} \delta. \tag{7.1.1}$$

PROOF. By the continuity of the Fourier transformation in $\mathcal{S}'$ and (6.9.3) we have

$$\widehat{T} = \sum_{\nu \in \mathbb{Z}} e^{i\nu(\cdot)}.$$

Let $\widehat{T}|_{(-\pi+2\nu\pi,\pi+2\nu\pi)}$ and $\widehat{T}|_{(0+2\pi\nu,2\pi+2\nu\pi)}$ be the restrictions of $\widehat{T}$ to the open sets $(-\pi + 2\nu\pi, \pi + 2\nu\pi)$ and $(0 + 2\pi\nu, 2\pi + 2\nu\pi)$, respectively. We shall now show that $\widehat{T}|_{(-\pi+2\nu\pi,\pi+2\nu\pi)} = 2\pi\tau_{2\pi\nu}\delta$ and $\widehat{T}|_{(0+2\pi\nu,2\pi+2\nu\pi)} = 0$. We shall only prove the case when $\nu = 0$. The proof in the rest cases runs in the same manner. Put $S_\nu(x) = \sum_{|\mu| \le \nu} e^{i\mu x}$. It is easy to check that

$$S_\nu(x) = \frac{\sin(\nu + \frac{1}{2})x}{\sin \frac{x}{2}}.$$

and $\int_{-\pi}^{\pi} S_\nu(x)dx = 2\pi$. Suppose that $\varphi \in \mathcal{D}(-\pi, \pi)$. We have to show that

$$\lim_{\nu \to \infty} \frac{1}{2\pi} \int_{-\pi}^{\pi} S_\nu(x)(\varphi(x) - \varphi(0))\, dx = 0.$$

In order to do it we only need to show that

$$\lim_{\nu \to \infty} \int_{-\pi}^{\pi} S_\nu(x)\varphi(x)\, dx = 0$$

if $\varphi(0) = 0$. In this case, the function $\varphi$ can be written as follows $\varphi(x) = x\psi(x)$, where $\psi$ is in $\mathcal{D}(-\pi, \pi)$. Further, note that

$$\int_{-\pi}^{\pi} \frac{\sin(\nu + \frac{1}{2})x}{\sin \frac{x}{2}}\varphi(x)\, dx = \int_{-\pi}^{\pi} \frac{x \cos \frac{x}{2}}{\sin \frac{x}{2}}\psi(x) \sin \nu x\, dx + \int_{-\pi}^{\pi} \varphi(x) \cos \nu x\, dx.$$

Since the function $f$, $f(x) = \frac{x \cos \frac{x}{2}}{\sin \frac{x}{2}} \psi(x)$ is continuous, therefore

$$\lim_{\nu \to \infty} \int_{-\pi}^{\pi} S_\nu(x)\varphi(x)\,dx = 0.$$

when $\varphi \in \mathcal{D}(-\pi, \pi)$ and $\varphi(0) = 0$. In the same manner we can see that

$$\lim_{\nu \to \infty} \int_0^{2\pi} \frac{\sin(\nu + \frac{1}{2})x}{\sin \frac{x}{2}} \varphi(x)\,dx = 0$$

if $\varphi \in \mathcal{D}(0, 2\pi)$. From this, by Theorem 2.3.2, we obtain (7.1.1). Thus, the proof is finished.          $\square$

(7.1.1) is known as the Poisson summation formula.

## 7.2. Periodic distributions

The following lemma will be needed in our considerations.

LEMMA 7.2.1. *There exists a function $\psi$ in $\mathcal{D}(-\frac{3}{4}, \frac{3}{4})$ such that $0 \le \psi(x) \le 1$, $\psi(x) = 1$ for $x \in [-\frac{1}{2}, \frac{1}{2}]$ and*

$$\sum_{\nu \in \mathbb{Z}} \psi(\nu - x) = 1$$

*for each $x \in \mathbb{R}$.*

PROOF. In view of Lemma 2.1.1 there exists a function $\theta \in \mathcal{D}(-\frac{3}{4}, \frac{3}{4})$ such that $0 \le \theta(x) \le 1$ and $\theta(x) = 1$ for $x \in [-\frac{1}{2}, \frac{1}{2}]$. Put

$$\omega(x) := \sum_{\nu \in \mathbb{Z}} \theta(\nu - x) \ge 1$$

and

$$\psi(x) := \frac{\theta(x)}{\omega(x)}.$$

It is easily seen that $\sum_{\nu \in \mathbb{Z}} \psi(\nu - x) = 1$ for each $x \in \mathbb{R}$. This finishes the proof.   $\square$

DEFINITION 7.2.1. We say that a distribution $T$ is $\alpha$-periodic, $\alpha > 0$ if $\tau_\alpha T = T$.

Having disposed of this preliminary step, we can prove the following

THEOREM 7.2.1. *Every 1-periodic distribution can be represented as follows*

$$T = \sum_{\nu \in \mathbb{Z}} c_\nu e^{i2\pi\nu(\cdot)} \tag{7.2.1}$$

*in $\mathcal{S}'$, where $c_\nu = T\left(\psi e^{-i2\pi\nu(\cdot)}\right)$, $\psi$ is as in Lemma 7.2.1. Moreover, there exist $M > 0$ and $m \in \mathbb{N}$ such that $|c_\nu| \le M|\nu|^m$ for $\nu \in \mathbb{Z} - \{0\}$.*

PROOF. Equality (7.1.1) can be written as follows

$$\sum_{\nu \in \mathbb{Z}} \widehat{\varphi}(\nu) = 2\pi \sum_{\nu \in \mathbb{Z}} \varphi(2\pi\nu) \tag{7.2.2}$$

for $\varphi \in \mathcal{S}$. Replacing $\varphi$ by $\varphi e^{ix(\cdot)}$ in (7.2.2), by (6.9.4) we get

$$\sum_{\nu \in \mathbb{Z}} \widehat{\varphi}(x + \nu) = 2\pi \sum_{\nu \in \mathbb{Z}} \varphi(2\pi\nu) e^{i2\pi\nu x}. \tag{7.2.3}$$

Let $\psi$ be as in Lemma 7.2.1, then

$$\psi(x) \sum_{\nu \in \mathbb{Z}} \widehat{\varphi}(x + \nu) = 2\pi \psi(x) \sum_{\nu \in \mathbb{Z}} \varphi(2\pi\nu) e^{i2\pi\nu x}.$$

We shall now show that the series

$$2\pi \sum_{\nu \in \mathbb{Z}} \varphi(2\pi\nu) \psi e^{i2\pi\nu(\cdot)} \tag{7.2.4}$$

is convergent in $\mathcal{D}$. Indeed, note that

$$\left( \psi e^{i2\pi\nu(\cdot)} \right)^{(k)} = \sum_{j=0}^{k} \binom{k}{j} (i2\pi\nu)^{k-j} \psi^{(j)} e^{i2\pi\nu(\cdot)}.$$

We only need to show that the series

$$\sum_{\nu \in \mathbb{Z}} \varphi(2\pi\nu)(i2\pi\nu)^{k-j} \psi^{(j)} e^{i2\pi\nu(\cdot)}, \quad j = 0, \ldots, k \tag{7.2.5}$$

is uniformly convergent in $\mathbb{R}$. This fact is evident because $\varphi$ is in $\mathcal{S}$. From this we infer that

$$T \left( \sum_{\nu \in \mathbb{Z}} \psi \widehat{\varphi}(\cdot + \nu) \right) = 2\pi \sum_{\nu \in \mathbb{Z}} \varphi(2\pi\nu) T \left( \psi e^{i2\pi\nu(\cdot)} \right). \tag{7.2.6}$$

Put

$$\tilde{c}_\nu = T \left( \psi e^{i2\pi\nu(\cdot)} \right) = T\psi(e^{i2\pi\nu(\cdot)}).$$

Since $T\psi$ has a compact support, therefore by Theorem 6.10.1 there exist two constants $M > 0$ and $m \in \mathbb{N}$ such that

$$|\tilde{c}_\nu| \le M|\nu|^m \tag{7.2.7}$$

for $\nu \in \mathbb{Z} - \{0\}$. Let $\varphi, \varphi_\mu \in \mathcal{S}$, $\mu = 1, 2, \ldots$. Assume that $\varphi_\mu \to \varphi$ in $\mathcal{S}$. Then we have

$$(1 + (2\pi\nu)^2)^{k+1} |\varphi_\mu(2\pi\nu)| \le K_k$$

for $\nu, \mu = 1, 2, \ldots$. Further,

$$|\tilde{c}_\nu \varphi_\mu(2\pi\nu)| \le M \frac{|\nu|^k}{(1 + (2\pi\nu)^2)^{k+1}} K_k \le M K_k \frac{1}{1 + (2\pi\nu)^2}.$$

This implies that the series $\sum_{\nu \in \mathbb{Z}} \tilde{c}_\nu \varphi_\mu(2\pi\nu)$ is uniformly convergent with respect to $\mu \in \mathbb{N}$. Therefore

$$\lim_{\mu \to \infty} \sum_{\nu \in \mathbb{Z}} T \left( \psi e^{i2\pi\nu(\cdot)} \right) \varphi_\mu(2\pi\nu) = \sum_{\nu \in \mathbb{Z}} T \left( \psi e^{i2\pi\nu(\cdot)} \right) \varphi(2\pi\nu). \tag{7.2.8}$$

Now, suppose that $\widehat{\varphi}_\mu$ is in $\mathcal{D}$ and that $\widehat{\varphi}_\mu$ converges to $\widehat{\varphi}$ in $\mathcal{S}$. Of course, $\varphi_\mu$ also converges to $\varphi$ in $\mathcal{S}$. Note that if $\widehat{\varphi}$ is in $\mathcal{D}$, then

$$T \left( \sum_{\nu \in \mathbb{Z}} \psi \widehat{\varphi}(\cdot + \nu) \right) = \sum_{\nu \in \mathbb{Z}} T (\psi \widehat{\varphi}(\cdot + \nu)).$$

Since $T$ is a 1-periodic distribution and $\psi \in \mathcal{D}$, therefore

$$T \left( \sum_{\nu \in \mathbb{Z}} \psi \widehat{\varphi}(\cdot + \nu) \right) = \sum_{\nu \in \mathbb{Z}} T(\widehat{\varphi}\psi(\cdot - \nu)) = T \left( \sum_{\nu \in \mathbb{Z}} \widehat{\varphi}\psi(\cdot - \nu) \right) = T(\widehat{\varphi}).$$

From this, by (7.2.6)

$$T(\widehat{\varphi}_\mu) = 2\pi \sum_{\nu \in \mathbb{Z}} T\left(\psi e^{i2\pi\nu(\cdot)}\right) \varphi_\mu(2\pi\nu).$$

Taking into account (7.2.8) we get

$$T(\widehat{\varphi}) = 2\pi \sum_{\nu \in \mathbb{Z}} \tilde{c}_\nu \varphi(2\pi\nu).$$

This equality can be written as follows

$$\widehat{T} = 2\pi \sum_{\nu \in \mathbb{Z}} \tilde{c}_\nu \tau_{2\pi\nu} \delta. \tag{7.2.9}$$

Since $\mathcal{F}1 = 2\pi\delta$, therefore by (6.9.4) we have

$$T = \sum_{\nu \in \mathbb{Z}} \tilde{c}_\nu e^{-i2\pi\nu(\cdot)}.$$

Replacing $-\nu$ by $\nu$ in this equality we obtain

$$T = \sum_{\nu \in \mathbb{Z}} c_\nu e^{i2\pi\nu(\cdot)}, \quad c_\nu = T(\psi e^{-i2\pi\nu(\cdot)}). \tag{7.2.10}$$

From (7.2.9) we infer that $c_\nu$ does not depend on $\psi$. This statement finishes the proof. $\qquad\square$

REMARK 7.2.1. If $T = f$, where $f$ is a 1-periodic continuous function, then $c_\nu = \int_{-\frac{1}{2}}^{\frac{1}{2}} f(x)e^{-2\pi\nu x}\,dx$. Indeed, we have

$$c_\nu = \int_{\mathbb{R}} f(x)\psi(x)e^{-i2\pi\nu x}\,dx = \sum_{\nu \in \mathbb{Z}} \int_{\nu-\frac{1}{2}}^{\nu+\frac{1}{2}} f(x)\psi(x)e^{-i2\pi\nu x}\,dx$$

$$= \sum_{\nu \in \mathbb{Z}} \int_{-\frac{1}{2}}^{\frac{1}{2}} f(\omega)\psi(\omega+\nu)e^{-i2\pi\nu\omega}\,d\omega = \int_{-\frac{1}{2}}^{\frac{1}{2}} f(x)e^{-i2\pi\nu x}\,dx.$$

From Theorem 7.2.1 it follows the following

COROLLARY 7.2.1. *Every periodic distribution is tempered.*

We shall now present the inverse theorem to the previous theorem.

THEOREM 7.2.2. *If there exist constants $M > 0$ and $m \in \mathbb{N}$ such that $|c_\nu| < M|\nu|^m$ for each $\nu \in \mathbb{Z} - \{0\}$, then the series*

$$\sum_{\nu \in \mathbb{Z}} c_\nu e^{i2\pi\nu(\cdot)} \tag{7.2.11}$$

*is convergent in $\mathcal{S}'$. Its sum is 1-periodic distribution.*

PROOF. Let

$$S_\nu = \sum_{|\mu| \le \nu} c_\mu e^{i2\pi\mu(\cdot)}.$$

Note that

$$\int_{\mathbb{R}} S_\nu \varphi = \sum_{|\mu| \le \nu} c_\mu \widehat{\varphi}(2\pi\mu)$$

for $\varphi \in \mathcal{S}$. Taking into account that $\hat{\varphi}$ is also in $\mathcal{S}$ by the assumption we conclude that the series $\sum_{\nu \in \mathbb{Z}} c_\nu \hat{\varphi}(2\pi\nu)$ is convergent. This means that the series (7.2.11) is convergent in $\mathcal{S}'$. It is easily seen that the mapping $\mathcal{S}' \ni T \to \tau_h T \in \mathcal{S}'$ is continuous for each $h \in \mathbb{R}$ (see Example 1.8.4). An immediate consequence of this remark are the following equalities

$$\tau_1 \sum_{\nu \in \mathbb{Z}} c_\nu e^{i2\pi\nu(\cdot)} = \sum_{\nu \in \mathbb{Z}} c_\nu \tau_1 e^{i2\pi\nu(\cdot)} = \sum_{\nu \in \mathbb{Z}} c_\nu e^{i2\pi\nu(\cdot)}.$$

Therefore, the distribution $T$, $T = \sum_{\nu \in \mathbb{Z}} c_\nu e^{i2\pi\nu(\cdot)}$ is a 1-periodic distribution. This statement finishes the proof. $\qquad\square$

## 7.3. The spaces $\mathcal{A}$ and $\mathcal{A}'$

We start with the presentation of some information about the Wiener orthogonal system. The Wiener functions

$$\rho_\nu(t) = \frac{1}{\sqrt{2\pi}} \frac{(-it - \frac{1}{2})^\nu}{(-it + \frac{1}{2})^{\nu+1}}, \quad \nu \in \mathbb{Z}, \ t \in \mathbb{R} \qquad (7.3.1)$$

constitute a complete orthonormal system in $L^2(\mathbb{R})$. The function $\rho_\nu$ is an eigenfunction of the differential operator $\mathcal{N}$,

$$\mathcal{N}u(t) := i(t - \frac{1}{2}i) \frac{d}{dt} \left( (t + \frac{1}{2}i)u(t) \right)$$

corresponding to the eigenvalue $-\nu$, $\nu \in \mathbb{Z}$. We shall now define a vector space $\mathcal{A}$ of smooth functions on the real line, which is very important in applications.

DEFINITION 7.3.1. We say that a smooth function $\varphi$ is in $\mathcal{A}$ if $\mathcal{N}^k\varphi \in L^2(\mathbb{R})$ and $\langle \mathcal{N}^k\varphi, \rho_\nu \rangle = \langle \varphi, \mathcal{N}^k\rho_\nu \rangle$ for each $k \in \mathbb{N}$ and $\nu \in \mathbb{Z}$ (see [27]).

Set by definition

$$\|\varphi\|_k := \|\mathcal{N}^k\varphi\|_{L^2}.$$

One can show that the set $\mathcal{A}$ is a complete vector space with the system of the seminorms $(\| \cdot \|_k, \ k \in \mathbb{N})$ ([27]). Because of identity

$$\langle (\mathcal{N}^2 + 1)^k\varphi, \varphi \rangle = \sum_{\nu=0}^{k} \binom{k}{\nu} \|\varphi\|_\nu^2$$

the system $(\| \cdot \|_k, \ k \in \mathbb{N})$ of seminorms $\| \cdot \|_k$ is equivalent to the following system $(\||| \cdot |||_k, \ k \in \mathbb{N})$ of norms, $\||\varphi\||_k^2 = \langle (\mathcal{N}^2 + 1)^k\varphi, \varphi \rangle$. Of course,

$$\||\varphi\||_0 \le \||\varphi\||_1 \le \cdots \le \||\varphi\||_k \le \ldots, \quad k \in \mathbb{N}. \qquad (7.3.2)$$

We shall now show that the norms $\||| \cdot |||_k$ are compatible in the following sense ([9]):

If a sequence $(\varphi_\nu)$, $\varphi_\nu \in \mathcal{A}$, converges to zero with respect to the norm $\||| \cdot |||_m$ and is a Cauchy sequence in the norm $\||| \cdot |||_n$, $m \le n$, then it converges to zero in the norm $\||| \cdot |||_n$.

Without loss of generality, one can assume that $m = 0$. Then we have

$$\begin{aligned}
|||\varphi_\nu|||_n^2 &= \langle (\mathcal{N}^2 + 1)^n \varphi_\nu, \varphi_\nu \rangle \\
&= \langle (\mathcal{N}^2 + 1)^n (\varphi_\nu - \varphi_\mu), \varphi_\nu \rangle + \langle (\mathcal{N}^2 + 1)^n \varphi_\mu, \varphi_\nu \rangle \\
&\leq \langle (\mathcal{N}^2 + 1)^n (\varphi_\nu - \varphi_\mu), \varphi_\nu - \varphi_\mu \rangle^{\frac{1}{2}} \langle (\mathcal{N}^2 + 1)^n \varphi_\nu, \varphi_\nu \rangle^{\frac{1}{2}} \\
&\quad + \|(\mathcal{N}^2 + 1)^n \varphi_\mu\|_{L^2} \|\varphi_\nu\|_{L^2} \\
&= |||\varphi_\nu - \varphi_\mu|||_n |||\varphi_\nu|||_n + \|(\mathcal{N}^2 + 1)^n \varphi_\mu\|_{L^2} \|\varphi_\nu\|_{L^2}.
\end{aligned}$$

For every $\epsilon > 0$ we can choose $M > 0$ and a positive integer $\nu_0$ so that

$$|||\varphi_\nu|||_n^2 \leq \epsilon M + \epsilon \|(\mathcal{N}^2 + 1)^n \varphi_\mu\|_{L^2}$$

for $\mu, \nu > \nu_0$. This implies that the sequence $(\varphi_\nu)$ converges to zero in the norm $||| \cdot |||_n$. Therefore the norms $||| \cdot |||_k$, $k \in \mathbb{N}$ are compatible.

Let $\mathcal{A}_k$ be the completion of $\mathcal{A}$ with respect to the norm $||| \cdot |||_k$. Every element $\tilde{f}$ in $\mathcal{A}_k$ is defined by the Cauchy sequence $(\varphi_\nu)$, $\varphi_\nu \in \mathcal{A}$, $\nu = 1, 2 \ldots$, with respect to the norm $||| \cdot |||_k$. According to (7.3.2) this sequence is also a Cauchy sequence with respect to the norm $\| \cdot \|_{L^2} = ||| \cdot |||_0$ and therefore defines a function $f$ in $L^2(\mathbb{R})$. It is easy to verify that $f$ is uniquely determined by $\tilde{f}$. We shall now show that distinct elements $\tilde{f}_1$ and $\tilde{f}_2$ from $\mathcal{A}_k$ cannot be mapped into the same element $f \in L^2(\mathbb{R})$. It is sufficient to verify that a nonzero element $\tilde{\psi}$ in $\mathcal{A}_k$ cannot be carried into the zero element of $L^2(\mathbb{R})$. Indeed, assuming the contrary, for a sequence $(\varphi_\nu)$, $\varphi_\nu \in \mathcal{A}$, which is a Cauchy sequence with respect to the norm $||| \cdot |||_k$ and defines the element $\tilde{f} \neq 0$, we would have the following relations

$$\lim_{\nu \to \infty} |||\varphi_\nu|||_k > 0, \quad \lim_{\nu \to \infty} \|\varphi_\nu\|_{L^2} = 0.$$

It would contradict the compatibility of the norms $||| \cdot |||_k$ and $\| \cdot \|_{L^2}$.

This consideration indicates us that the completion $\mathcal{A}_k$ of $\mathcal{A}$ may be regarded as a subspace of $L^2(\mathbb{R})$. A function $\varphi \in L^2(\mathbb{R})$ is in $\mathcal{A}_k$ iff there exists a sequence $(\varphi_\nu)$, $\varphi_\nu \in \mathcal{A}$ such that $\varphi_\nu$ tends to $\varphi$ in $L^2$ as $\nu \to \infty$ and is a Cauchy sequence with respect to the norm $||| \cdot |||_k$. Of course, we have

$$L^2(\mathbb{R}) = \mathcal{A}_0 \supset \mathcal{A}_1 \supset \ldots \tag{7.3.3}$$

By continuity, one can extend the norms $||| \cdot |||_k$ from $\mathcal{A}$ onto $\mathcal{A}_k$ in the following way: assuming that $\varphi$ is in $\mathcal{A}_k$, the sequence $(\varphi_\nu)$, $\varphi_\nu \in \mathcal{A}$ converges to $\varphi$ in $L^2$ and $(\varphi_\nu)$ is a Cauchy sequence with respect to the norm $||| \cdot |||_k$, we take

$$|||\varphi|||_k = \lim_{\nu \to \infty} |||\varphi_\nu|||_k.$$

Analogously, the bilinear form $(\cdot, \cdot)_k := \langle (\mathcal{N}^2 + 1) \cdot, \cdot \rangle$ can be extended by continuity from $\mathcal{A} \times \mathcal{A}$ onto $\mathcal{A}_k \times \mathcal{A}_k$ by taking

$$\begin{aligned}
(\varphi, \psi)_k := \lim_{\nu \to \infty} \frac{1}{4} \Big( & (|||\varphi_\nu + \psi_\nu|||_k^2 - |||\varphi_\nu - \psi_\nu|||_k^2) \\
& - i(|||\varphi_\nu + i\psi_\nu|||_k^2 - |||\varphi_\nu - i\psi_\nu|||_k^2) \Big), \quad (7.3.4)
\end{aligned}$$

where the sequences $(\varphi_\nu)$ and $(\psi_\nu)$ determine $\varphi$ and $\psi$ in $\mathcal{A}_k$. It is easy to check that

$$(\varphi, \psi)_k = \overline{(\psi, \varphi)_k}, \quad \varphi, \psi \in \mathcal{A}_k, \tag{7.3.5}$$

$$(\varphi, \varphi)_k \leq (\varphi, \varphi)_{k+1}, \quad \varphi \in \mathcal{A}_{k+1}, \tag{7.3.6}$$

$$(\varphi, \psi)_k = \langle (\mathcal{N}^2 + 1)^k \varphi, \psi \rangle, \quad \varphi \in \mathcal{A}, \ \psi \in \mathcal{A}_k. \tag{7.3.7}$$

The vector space $\mathcal{A}_k$ under the bilinear form $(\cdot, \cdot)_k$ is a Hilbert space. Every continuous linear form $\Lambda$ on $\mathcal{A}$ can be extended by continuity on some space $\mathcal{A}_k$.

We shall now construct an orthonormal system in $\mathcal{A}_k$, $k = 1, 2, \ldots$. In order to do this we put

$$\rho_{\nu k} = (\nu^2 + 1)^{-\frac{k}{2}} \rho_\nu.$$

THEOREM 7.3.1. *The functions $\rho_{\nu k}$, $\nu \in \mathbb{Z}$ constitute a complete orthonormal system in $\mathcal{A}_k$.*

PROOF. From (7.3.7), it follows that the functions $\rho_{\nu k}$, $\nu \in \mathbb{Z}$ constitute an orthonormal system in $\mathcal{A}_k$ with respect to the bilinear form $(\cdot \cdot)_k$. It remains to prove that the system $(\rho_{\nu k})$ is complete in $\mathcal{A}_k$. Indeed, let $\varphi$ be in $\mathcal{A}_k$, then

$$\varphi = \sum_{\nu \in \mathbb{Z}} (\varphi, \rho_\nu) \rho_\nu$$

is in $L^2(\mathbb{R})$. The series $\sum_{\nu \in \mathbb{Z}} (\varphi, \rho_{\nu k})_k \rho_{\nu k}$ may be regarded as the Fourier series of $\varphi$ under the orthonormal system $(\rho_{\nu k})$. Of course,

$$\sum_{\nu \in \mathbb{Z}} |(\varphi, \rho_{\nu k})_k|^2 \leq |||\varphi|||_k^2.$$

Since $\mathcal{A}_k$ is a Hilbert space with respect to the norm $||| \cdot |||_k$, therefore there exists $\psi$ in $\mathcal{A}_k$ such that $\psi = \sum_{\nu \in \mathbb{Z}} (\varphi, \rho_{\nu k})_k \rho_{\nu k}$ in $\mathcal{A}_k$. Because of (7.3.7) we have

$$(\varphi, \rho_{\nu k})_k = (\nu^2 + 1)^{\frac{k}{2}} (\varphi, \rho_\nu). \tag{7.3.8}$$

Hence we get $\psi = \sum_{\nu \in \mathbb{Z}} (\varphi, \rho_\nu) \rho_\nu$ in $\mathcal{A}_k$. Since the norms $||| \cdot |||_k$ and $\| \cdot \|_{L^2}$ are compatible, therefore $\varphi = \psi$ in $\mathcal{A}_k$. Thus, the proof of the theorem is finished. $\square$

Because of $\sum_{\nu \in \mathbb{Z}} |(\varphi, \rho_{\nu k})_k|^2 \leq |||\varphi|||_k^2$, by (7.3.8) we have

$$\sum_{\nu \in \mathbb{Z}} (\nu^2 + 1)^k |c_\nu|^2 < \infty, \quad c_\nu = (\varphi, \rho_\nu). \tag{7.3.9}$$

Conversely, assume that (7.3.9) is fulfilled, then by the completeness of $\mathcal{A}_k$ we infer that the series $\sum_{\nu \in \mathbb{Z}} (\nu^2 + 1)^{\frac{k}{2}} c_\nu \rho_{\nu k}$ converges to some $\varphi$ in $\mathcal{A}_k$, therefore $\varphi$ is in $\mathcal{A}_k$.

We have proved the fact which can be stated as a separate

THEOREM 7.3.2. *A function $\varphi$ belonging to $L^2(\mathbb{R})$ is in $\mathcal{A}_k$ iff (7.3.9) holds.*

COROLLARY 7.3.1. *A function $\varphi$ belonging to $L^2(\mathbb{R})$ is in $\mathcal{A}$ iff (7.3.9) holds for each $k \in \mathbb{N}$.*

We shall now describe the Fourier expansion of the linear continuous forms being in the dual space $\mathcal{A}_{-k}$ of $\mathcal{A}_k$. Let $\mathcal{A}'$ denote the vector space of all continuous linear form over $\mathcal{A}$. Every element $\Lambda$ of $\mathcal{A}'$ can be extended on some space $\mathcal{A}_k$. In this case we say that $\Lambda$ is in $\mathcal{A}_{-k}$.

THEOREM 7.3.3. *An element $\Lambda$ from $\mathcal{A}'$ is in $\mathcal{A}_{-k}$ iff*

$$\sum_{\nu \in \mathbb{Z}} \frac{|C_\nu|^2}{(\nu^2 + 1)^k} < \infty, \quad C_\nu = \Lambda(\overline{\rho}_\nu). \tag{7.3.10}$$

The number $\Lambda(\overline{\rho}_\nu)$ is called the Fourier coefficient of $\Lambda$.

PROOF. Let $\Lambda$ be in $\mathcal{A}_{-k}$, then by Riesz theorem there exists a function $\psi$ in $\mathcal{A}_k$ such that

$$\Lambda(\overline{\varphi}) = (\psi, \varphi)_k \tag{7.3.11}$$

for each $\varphi \in \mathcal{A}_k$. The function $\psi$ has the following Fourier representation

$$\psi = \sum_{\nu \in \mathbb{Z}} c_{\nu k} \rho_{\nu k}, \quad c_{\nu k} = (\psi, \rho_{\nu k})_k \tag{7.3.12}$$

and

$$\|\Lambda\|^2_{\mathcal{A}_{-k}} = \||\psi\||^2_k = \sum_{\nu \in \mathbb{Z}} |c_{\nu k}|^2. \tag{7.3.13}$$

In view of (7.3.8) we obtain

$$\Lambda(\overline{\varphi}) = (\sum_{\nu \in \mathbb{Z}} c_{\nu k} \rho_{\nu k}, \varphi)_k = \sum_{\nu \in \mathbb{Z}} c_{\nu k} (\nu^2 + 1)^{\frac{k}{2}} (\rho_\nu, \varphi).$$

Putting $\varphi = \rho_\mu$ in the previous equality we get

$$\Lambda(\overline{\rho}_\mu) = (\mu^2 + 1)^{\frac{k}{2}} c_{\mu k}. \tag{7.3.14}$$

Finally we have

$$\Lambda(\overline{\varphi}) = \sum_{\nu \in \mathbb{Z}} \Lambda(\overline{\rho}_\nu)(\rho_\nu, \varphi), \quad \varphi \in \mathcal{A}_k. \tag{7.3.15}$$

From (7.3.13) and (7.3.14) it follows the following equality

$$\||\psi\||^2_k = \sum_{\nu \in \mathbb{Z}} \frac{|C_\nu|^2}{(\nu^2 + 1)^k}. \tag{7.3.16}$$

Therefore (7.3.10) holds. It remains to prove that (7.3.10) implies the existence of a function $\psi$ in $\mathcal{A}_k$ such that (7.3.11) holds. It is easy to check that the function $\psi = \sum_{\nu \in \mathbb{Z}} \frac{C_\nu}{(\nu^2+1)^k} \rho_\nu$ has the required properties. $\qquad \square$

## 7.4. Cauchy transforms of elements of $\mathcal{A}'$

We begin by determining the Fourier coefficients of the function $\frac{1}{2\pi i} \frac{1}{(\cdot - z)}$, $\Im z \neq 0$ with respect to the Wiener function $\rho_\nu$, $\nu \in \mathbb{Z}$. First let's assume that $\nu = 0, 1, 2, \ldots$. We shall separately consider two cases: $\Im z < 0$ and $\Im z > 0$. Note that

$$\overline{\rho}_\nu(t) = \frac{1}{\sqrt{2\pi}} \frac{(it - \frac{1}{2})^\nu}{(it + \frac{1}{2})^{\nu+1}}$$

is holomorphic if $\Im t < \frac{1}{2}$. Suppose that $\Im z < 0$. Then, it is easy to check that

$$c_\nu \left( \frac{1}{2\pi i} \frac{1}{(\cdot - z)} \right) = \frac{1}{2\pi i} \int_{\mathbb{R}} \frac{1}{\sqrt{2\pi}} \frac{(it - \frac{1}{2})^\nu}{(it + \frac{1}{2})^{\nu+1}} \frac{1}{t - z} \, dt = \frac{-1}{\sqrt{2\pi}} \frac{(iz - \frac{1}{2})^\nu}{(iz + \frac{1}{2})^{\nu+1}}$$

and

$$\left| \frac{iz - \frac{1}{2}}{iz + \frac{1}{2}} \right| < 1. \tag{7.4.1}$$

On the other hand for $\Im z > 0$ the function $\frac{(it-\frac{1}{2})^\nu}{(it+\frac{1}{2})^{\nu+1}} \frac{1}{(t-z)}$ is holomorphic if $\Im t < 0$. This implies that

$$c_\nu \left( \frac{1}{2\pi i} \frac{1}{(\cdot - z)} \right) = 0.$$

In the same manner one can show that if $\nu = -1, -2, \ldots$, then

$$c_\nu \left( \frac{1}{2\pi i} \frac{1}{(\cdot - z)} \right) = \frac{1}{\sqrt{2\pi}} \frac{(iz - \frac{1}{2})^\nu}{(iz + \frac{1}{2})^{\nu+1}} \quad \text{if } \Im z > 0$$

and

$$c_\nu \left( \frac{1}{2\pi i} \frac{1}{(\cdot - z)} \right) = 0 \quad \text{if } \Im z < 0.$$

It is easily seen that

$$\left| \frac{iz - \frac{1}{2}}{iz + \frac{1}{2}} \right| > 1 \quad \text{if } \Im z > 0. \tag{7.4.2}$$

Thus, we obtain the following Fourier expansion of $\frac{1}{2\pi i} \frac{1}{(\cdot - z)}$ with respect to the Wiener functions

$$\frac{1}{2\pi i} \frac{1}{(\cdot - z)} = \frac{1}{\sqrt{2\pi}} \sum_{\nu=-\infty}^{-1} \frac{(iz - \frac{1}{2})^\nu}{(iz + \frac{1}{2})^{\nu+1}} \rho_\nu \quad \text{if } \Im z > 0, \tag{7.4.3}$$

$$\frac{1}{2\pi i} \frac{1}{(\cdot - z)} = -\frac{1}{\sqrt{2\pi}} \sum_{\nu=0}^{\infty} \frac{(iz - \frac{1}{2})^\nu}{(iz + \frac{1}{2})^{\nu+1}} \rho_\nu \quad \text{if } \Im z < 0. \tag{7.4.4}$$

From (7.4.2) and (7.4.1) it follows that series (7.4.3) and (7.4.4) converge in $\mathcal{A}$ if $\Im z > 0$ and $\Im z < 0$, respectively. Let $\Lambda$ be in $\mathcal{A}_{-k}$, then we get the following equalities

$$\Lambda \left( \frac{1}{2\pi i} \frac{1}{(\cdot - z)} \right) = \frac{1}{\sqrt{2\pi}} \sum_{\nu=-\infty}^{-1} \Lambda(\rho_\nu) \frac{(iz - \frac{1}{2})^\nu}{(iz + \frac{1}{2})^{\nu+1}} \quad \text{if } \Im z > 0 \tag{7.4.5}$$

$$\Lambda \left( \frac{1}{2\pi i} \frac{1}{(\cdot - z)} \right) = -\frac{1}{\sqrt{2\pi}} \sum_{\nu=0}^{\infty} \Lambda(\rho_\nu) \frac{(iz - \frac{1}{2})^\nu}{(iz + \frac{1}{2})^{\nu+1}} \quad \text{if } \Im z < 0 \tag{7.4.6}$$

This equalities are understood in the sense of $\mathcal{A}_{-k}$. According to (7.4.1) and (7.4.2) we have

$$\Pi^- \ni z \to v := \frac{iz - \frac{1}{2}}{iz + \frac{1}{2}} \in B_1(0)$$

and

$$\Pi^+ \ni z \to w := \frac{iz + \frac{1}{2}}{iz - \frac{1}{2}} \in B_1(0).$$

Since $\Lambda$ is in $\mathcal{A}_{-k}$, therefore (7.3.10) holds. This implies that the series

$$\frac{1}{\sqrt{2\pi}} \sum_{\nu=0}^{\infty} \Lambda(\rho_\nu) v^\nu \tag{7.4.7}$$

and

$$\frac{1}{\sqrt{2\pi}} \sum_{\mu=1}^{\infty} \Lambda(\rho_{-\mu}) w^\mu \tag{7.4.8}$$

are almost uniformly convergent in $B_1(0)$. Taking into account (7.4.1) and (7.4.2) we conclude that series (7.4.5) and (7.4.6) also almost uniformly converge in $\Pi^+$ and $\Pi^-$, respectively. Thus, we have obtained the following

THEOREM 7.4.1. *If* $\Lambda \in \mathcal{A}_{-k}$, *then the function*

$$C\Lambda(z) := \Lambda \left( \frac{1}{2\pi i} \frac{1}{(\cdot - z)} \right)$$

*given by formulas (7.4.5) and (7.4.6) is holomorphic in* $\Pi^- \cup \Pi^+$.

COROLLARY 7.4.1. *If* $\Lambda \in \mathcal{A}'$, *then the function* $C\Lambda$ *is holomorphic in* $\Pi^- \cup \Pi^+$.

DEFINITION 7.4.1. The function $C\Lambda$ defined by (7.4.5) and (7.4.6) is called the Cauchy transform of $\Lambda$.

### 7.5. The Wiener expansion of square integrable distributions

It is clear that $\mathcal{D} \subset \mathcal{S} \subset \mathcal{A}$ and that these embedings are continuous.

THEOREM 7.5.1. $\mathcal{A} \subset \mathcal{D}_{L^2}$.

PROOF. Let $\varphi$ be in $\mathcal{A}$. In accordance with Definition 7.3.1, $\varphi$ is in $L^2(\mathbb{R})$. Note that

$$|\rho_\nu(t)| = \frac{1}{\sqrt{2\pi}} \left( t^2 + \frac{1}{4} \right)^{-\frac{1}{2}}. \tag{7.5.1}$$

An easy computation shows that

$$\rho_\nu^{(1)} = -(\nu + 1)i\rho_{\nu+1} + (2\nu + 1)i\rho_\nu - \nu i\rho_{\nu-1}. \tag{7.5.2}$$

According to Corollary 7.3.1 the function $\varphi$ can be represented as follows

$$\varphi = \sum_{\nu \in \mathbb{Z}} c_\nu \rho_\nu, \quad c_\nu = \int_{\mathbb{R}} \varphi \bar{\rho}_\nu.$$

Formally differentiating this series we obtain

$$\sum_{\nu \in \mathbb{Z}} c_\nu \rho_\nu^{(1)} = -i \sum_{\nu \in \mathbb{Z}} (\nu + 1)c_\nu \rho_{\nu+1} + i \sum_{\nu \in \mathbb{Z}} (2\nu + 1)c_\nu \rho_\nu - i \sum_{\nu \in \mathbb{Z}} \nu c_\nu \rho_{\nu-1}.$$

By virtue of Corollary 7.3.1 it follows that the series $\sum_{\nu \in \mathbb{Z}} (\nu+1)c_\nu$, $\sum_{\nu \in \mathbb{Z}} (2\nu+1)c_\nu$ and $\sum_{\nu \in \mathbb{Z}} \nu c_\nu$ are absolutely convergent. Hence, by (7.5.1) $\varphi^{(1)}$ is in $L^2(\mathbb{R})$. In general, it follows by induction that $\varphi^{(k)} \in L^2(\mathbb{R})$ for each $k \in \mathbb{N}$. Thus, the proof is finished. □

Let $\Lambda$ be in $\mathcal{D}'_{L^2}(\mathbb{R})$. Denote by $\overset{\circ}{\Lambda}$ the restriction of $\Lambda$ to the set $\mathcal{A}$.

THEOREM 7.5.2. $\overset{\circ}{\Lambda}$ *is in* $\mathcal{A}'$.

PROOF. Because of Theorem 5.7.2, $C\Lambda(\cdot + i\epsilon) - C\Lambda(\cdot - i\epsilon)$ tends to $\Lambda$ in $\mathcal{D}'_{L^2}$. Therefore, in view of Theorem 7.5.1 we infer that $C\Lambda(\cdot + i\epsilon) - C\Lambda(\cdot - i\epsilon)$ converges to $\overset{\circ}{\Lambda}$ in $\mathcal{A}'$. Taking into account that $\mathcal{A}$ is complete we conclude by Theorem A.1 that $\overset{\circ}{\Lambda}$ belongs to $\mathcal{A}'$. □

COROLLARY 7.5.1. *If $\Lambda$ is in $\mathcal{D}'_{L^2}$, then*

$$\overset{\circ}{\Lambda} = \sum_{\nu \in \mathbb{Z}} c_\nu \rho_\nu$$

*in $\mathcal{A}'$, where $c_\nu = \overset{\circ}{\Lambda}(\overline{\rho}_\nu)$.*

Since

$$C\Lambda(x + iy) - C\Lambda(x - iy) = \Lambda(g_y(x - \cdot))$$

for $y > 0$, where $g_y(x) = \frac{1}{\pi} \frac{y}{x^2 + y^2}$ (compare with (5.6.1)), therefore

$$u(x, y) := \Lambda(g_y(x - \cdot)) = \frac{1}{\sqrt{2\pi}} \sum_{\nu = -\infty}^{-1} \Lambda(\rho_\nu) \frac{(i(x + iy) - \frac{1}{2})^\nu}{(i(x + iy) + \frac{1}{2})^{\nu+1}}$$

$$+ \frac{1}{\sqrt{2\pi}} \sum_{\nu = 0}^{\infty} \Lambda(\rho_\nu) \frac{(i(x - iy) - \frac{1}{2})^\nu}{(i(x - iy) + \frac{1}{2})^{\nu+1}}. \quad (7.5.3)$$

From this, for $\Lambda$ belonging to $\mathcal{D}'_{L^2}$ as an immediate consequence of Theorem 5.7.2 we obtain the following

THEOREM 7.5.3. *If $\Lambda$ is in $\mathcal{D}'_{L^2}$, then the function given by (7.5.3) fulfils the Laplace equation $\Delta u = 0$ in $\Pi^+$ and $\lim_{y \to 0^+} u(\cdot, y) = \Lambda$ in the sense of $\mathcal{D}'_{L^2}$ (compare with Theorem 5.6.4).*

# APPENDIX

# SEQUENTIAL COMPLETENESS OF SOME SPACES

Let $E$ be a vector space over the field of complex numbers. Let us equip this space with a family of norms $\| \cdot \|_m$ fulfilling the inequalities

$$\|\varphi\|_m \le \|\varphi\|_{m+1} \text{ for } \varphi \in E$$

and $m = 0, 1, 2 \ldots$.

A sequence $(\varphi_\nu)$ is called convergent to $\varphi$ in $E$ if for $m \in \mathbb{N}$, $\|\varphi_\nu - \varphi\|_m$ tends to zero as $\nu \to \infty$. If for $m \in \mathbb{N}$, $\|\varphi_\nu - \varphi_\mu\|_m$ goes to zero as $\nu$ and $\mu$ tend to infinity, then we say that the sequence $(\varphi_\nu)$ is a Cauchy sequence in $E$.

A vector space is said to be sequentially complete if every Cauchy sequence of elements of $E$ is convergent to an element of $E$. A vector space equipped with a family of norms $\| \cdot \|_m$, $m \in \mathbb{N}$, which is complete, is called a Fréchet space.

Denote by $E'$ the space of all linear continuous forms on $E$. We say that a sequence $(\Lambda_\nu)$, $\Lambda_\nu \in E'$ is a Cauchy sequence if for each $\varphi$ in $E$ the numerical sequence $(\Lambda_\nu(\varphi))$ is a Cauchy sequence in the usual sense. The space $E'$ is called sequentially complete if for every Cauchy sequence $(\Lambda_\nu)$ the function $\Lambda$ defined by

$$\Lambda(\varphi) = \lim_{\nu \to \infty} \Lambda_\nu(\varphi) \tag{A.1}$$

for $\varphi \in E$ is a linear continuous form on $E$. After the above preliminaries we are now in a position to prove the following

THEOREM A.1. *If $E$ is a Fréchet space, then its dual space $E'$ is sequentially complete.*

PROOF. It is easy to see that the function $\Lambda$ defined by (A.1) is a linear form on $E$. It remains to prove that $\Lambda$ is continuous. Let $(\varphi_\nu)$ be a sequence converging to zero. We must show that $\Lambda(\varphi_\nu) \to 0$. Let us assume the contrary. Then choosing, if necessary, a subsequence we can assume that $|\Lambda(\varphi_\nu)| \ge c > 0$. Now recall that convergence of a sequence $(\varphi_\nu)$ to zero in $E$ means that $\|\varphi_\nu\|_m \to 0$ for $m \in \mathbb{N}$. Again choosing a subsequence, we can assume that

$$\|\varphi_\nu\|_m \le \frac{1}{4^\nu}, \quad m = 0, 1, 2 \ldots, \nu. \tag{A.2}$$

Put by definition $\psi_\nu = 2^\nu \varphi_\nu$, $\nu \in \mathbb{N}$. Of course, the sequence $(\psi_\nu)$ is also convergent to zero in $E$, but

$$|\Lambda(\psi_\nu)| \to \infty.$$

We now define a subsequence $(\psi'_\nu)$ as follows: choose $\psi'_1$ so that $|\Lambda(\psi'_1)| > 1$. Now, since $\Lambda_\nu(\psi) \to \Lambda(\psi)$ for $\psi \in E$, we can choose $\Lambda'_1$ such that

$$|\Lambda'_1(\psi'_1)| > 1.$$

Now suppose we have chosen $\Lambda'_j$ and $\psi'_j$, $j = 1, 2, \ldots, \nu - 1$. We choose $\psi'_\nu$, $\nu > 1$ to be one of the $\psi_\nu$ with index so high that

$$|\Lambda'_k(\psi'_\nu)| < \frac{1}{2^{\nu-k}}, \quad k = 1, 2, \ldots, \nu - 1, \tag{A.3}$$

$$|\Lambda(\psi'_\nu)| > \sum_{j=1}^{\nu-1} |\Lambda(\psi'_j)| + \nu. \tag{A.4}$$

The first is possible, because the $\psi_\nu$ converge to zero in $E$ and $\Lambda_k$ are in $E'$. The second is possible, because $|\Lambda(\psi_\nu)| \to \infty$. Since $\Lambda_\nu(\psi) \to \Lambda(\psi)$ for $\psi \in E$, we can choose $\Lambda'_\nu$ from the $(\Lambda_\nu)$ sequence such that

$$|\Lambda'_\nu(\psi'_\nu)| > \sum_{j=1}^{\nu-1} |\Lambda'_\nu(\psi'_j)| + \nu. \tag{A.5}$$

This way we can go and construct the new infinite sequences $(\psi'_\nu)$ and $(\Lambda'_\nu)$. In view of (A.2) the sequence $(s_\nu)$, $s_\nu = \sum_{j=1}^{\nu} \psi'_j$ is a Cauchy sequence in $E$. $E$ is a Fréchet space, therefore there exists a $\psi$ in $E$ such that $\psi = \sum_{j=1}^{\infty} \psi'_j$. Further, for fixed $\nu$ we have

$$\Lambda'_\nu(\psi) = \sum_{j=1}^{\nu-1} \Lambda'_\nu(\psi'_j) + \Lambda'_\nu(\psi'_\nu) + \sum_{j=\nu+1}^{\infty} \Lambda'_\nu(\psi'_j).$$

Taking into account (A.3) and (A.5) we obtain

$$|\Lambda'_\nu(\psi)| \geq |\Lambda'_\nu(\psi'_\nu)| - \left| \sum_{j=1}^{\nu-1} \Lambda'_\nu(\psi'_j) + \sum_{j=\nu+1}^{\infty} \Lambda'_\nu(\psi'_j) \right|$$

$$\geq |\Lambda'_\nu(\psi'_\nu)| - \left( \sum_{j=1}^{\nu-1} |\Lambda'_\nu(\psi'_j)| + \sum_{j=\nu+1}^{\infty} |\Lambda'_\nu(\psi'_j)| \right)$$

$$\geq \nu - 1.$$

Therefore $|\Lambda'_\nu(\psi)| \to \infty$ as $\nu \to \infty$. But this contradicts $\lim_{\nu \to \infty} \Lambda'_\nu(\psi) = \Lambda(\psi)$. Consequently, $\Lambda(\varphi_\nu) \to 0$ and the linear form is therefore continuous. Thus, the proof of the theorem is finished. $\qquad\square$

We shall now deal with a more complicated problem. Let $\mathcal{E} = \{E_\gamma : \gamma \in \Gamma\}$ be a family of vector subspaces of the vector space $E$, such that $E = \bigcup_{\gamma \in \Gamma} E_\gamma$. Suppose that each of these spaces is a Fréchet space. We introduce a convergence in the space $E$.

A sequence $(\varphi_\nu)$ is called convergent in $E$ if it satisfies the following conditions:

$$(\alpha) \begin{cases} (i) & \varphi_\nu \in E_\gamma, \ \nu \in \mathbb{N} \text{ for some } \gamma \in \Gamma, \\ (ii) & (\varphi_\nu) \text{ is a Cauchy sequence in } E_\gamma. \end{cases}$$

Sequences fulfilling $(i)$ and $(ii)$ will be called $\alpha$-convergent.

Let $E'$ denote the set of all linear forms defined on $E$ and continuous with respect to the $\alpha$-convergence. We understand Cauchy sequences and the sequential completeness as in the above case. We shall now formulate the following

THEOREM A.2. *If $E$ is a vector space equipped with an $\alpha$-convergence, then the space $E'$ of all linear forms defined on $E$ and continuous with respect to this convergence is sequentially complete.*

Since the proof of this theorem is much the same as the proof of the previous theorem we will not give it here.

EXAMPLE A.1. Let $\Omega$ be an open set in $\mathbb{R}^n$ and $K \subset \Omega$ be a compact set. Denote by $\mathcal{D}(K)$ the set of all $\varphi$ belonging to $\mathcal{D}(\Omega)$ such that $\operatorname{supp}\varphi \subset K$. Of course, $\mathcal{D}(K)$ is a vector subspace of $\mathcal{D}(\Omega)$. It is clear that

$$\mathcal{D}(\Omega) = \bigcup_{K \subset \Omega} \mathcal{D}(K).$$

In accordance with Definition 1.1.1, $\mathcal{D}(K)$ is a Fréchet space and the convergence in $\mathcal{D}(\Omega)$ is $\alpha$-convergence. Therefore $\mathcal{D}'(\Omega)$ is sequentially complete. In practice, $K$ is always the closure of some open set.

We shall now collect some information concerning bilinear forms. Suppose that $E_1$ and $E_2$ are two vector spaces over the complex field $\mathbb{C}$ and that $\langle \cdot, \cdot \rangle$ is a mapping of $E_1 \times E_2$ into $\mathbb{C}$. Then, for each fixed $\psi \in E_2$, $\langle \cdot, \cdot \rangle$ defines a partial mapping $\langle \cdot, \psi \rangle$ of $E_1$ into $\mathbb{C}$; and similarly, $\langle \varphi, \cdot \rangle$ defines a partial mapping of $E_2$ into $\mathbb{C}$. If all the partial mappings $\langle \cdot, \psi \rangle$ and $\langle \varphi, \cdot \rangle$ are linear, then $\langle \cdot, \cdot \rangle$ is called a bilinear form on $E_1 \times E_2$. Let us equip these spaces with countable families of norms $\| \cdot \|_{m,1}$ and $\| \cdot \|_{m,2}$ fulfilling the inequalities

$$\|\varphi\|_{m,1} \le \|\varphi\|_{m+1,1}$$

and

$$\|\psi\|_{m,2} \le \|\psi\|_{m+1,2}$$

for $\varphi \in E_1$ and $\psi \in E_2$, respectively. Moreover, we assume that $E_2$ is complete. If the partial linear mappings $\langle \varphi, \cdot \rangle$ and $\langle \cdot, \psi \rangle$ are continuous, then we say that the bilinear form $\langle \cdot, \cdot \rangle$ is separately continuous. We are able to prove the following

THEOREM A.3. *Assume that the spaces $E_1$ and $E_2$ have the properties listed above. If $\langle \cdot, \cdot \rangle$ is a bilinear form on $E_1 \times E_2$ separately continuous, then it is continuous.*

PROOF. We have to show that there exist a constant $M > 0$ and nonnegative integers $m$ and $n$ such that

$$|\langle \varphi, \psi \rangle| \le M \|\varphi\|_{m,1} \|\psi\|_{n,2}$$

for $\varphi \in E_1$ and $\psi \in E_2$. On the contrary, suppose that our bilinear form is not continuous, therefore for every $M > 0$ and $m, n \in \mathbb{N}$ there exist $\tilde{\varphi}_{m,n,M}$ and $\tilde{\psi}_{m,n,M}$ such that

$$|\langle \tilde{\varphi}_{m,n,M}, \tilde{\psi}_{m,n,M} \rangle| > M \|\tilde{\varphi}_{m,n,M}\|_{m,1} \|\tilde{\psi}_{m,n,M}\|_{n,2}.$$

Set $m = n$, $M = n^2$, then we have

$$|\langle \tilde{\varphi}_{n,n,n^2}, \tilde{\psi}_{n,n,n^2} \rangle| > n^2 \|\tilde{\varphi}_{n,n,n^2}\|_{n,1} \|\tilde{\psi}_{n,n,n^2}\|_{n,2}.$$

For simplicity of notation we put

$$\overline{\varphi}_n := \frac{1}{n} \frac{\tilde{\varphi}_{n,n,n^2}}{\|\tilde{\varphi}_{n,n,n^2}\|}, \quad \overline{\psi}_n := \frac{1}{n} \frac{\tilde{\psi}_{n,n,n^2}}{\|\tilde{\psi}_{n,n,n^2}\|}.$$

Therefore $|\langle \overline{\varphi}_n, \overline{\psi}_n \rangle| > 1$. We shall now show that the $\overline{\varphi}_n$ converge to zero in $E_1$ and the $\overline{\psi}_n$ converge to zero in $E_2$. Note that

$$\|\overline{\varphi}_{n+k}\|_{k,1} \leq \|\overline{\varphi}_{n+k}\|_{k+n,1} = \frac{1}{n+k}, \quad k = 0, \ldots, \ n = 1, 2, \ldots$$

This implies that $\|\overline{\varphi}_{n+k}\|_{k,1}$ tends to 0 as $n \to \infty$. From this we infer that $\overline{\varphi}_n, \overline{\psi}_n$ tend to zero in $E_1$ and $E_2$, respectively. Then, choosing if necessary a subsequence, we may assume that

$$\|\overline{\varphi}_\nu\|_{k,1} < \frac{1}{4^\nu} \quad \text{and} \quad \|\overline{\psi}_\nu\|_{k,2} < \frac{1}{4^\nu}$$

for $k = 0, 1, \ldots, \nu$. Put $\varphi_\nu = 2^\nu \overline{\varphi}_\nu$ and $\psi_\nu = 2^\nu \overline{\psi}_\nu$. Then the sequence $\langle \varphi_\nu, \psi_\nu \rangle$ converges to $(0,0)$ in $E_1 \times E_2$. But

$$|\langle \varphi_\nu, \psi_\nu \rangle| > 2^{2\nu}, \quad \nu = 0, 1, \ldots \tag{A.6}$$

Of course, $|\langle \varphi_1, \psi_1 \rangle| > 1$. Put $\varphi_1' = \varphi_1$ and $\psi_1' = \psi_1$. By the assumption the function $\langle \varphi_1', \cdot \rangle$ is continuous on $E_2$ and (A.6) holds. Moreover, the $\psi_\nu$ converge to zero in $E_2$. Hence it follows that one can choose $\psi_2'$ from the sequence $(\psi_\nu)$ and $\varphi_2'$ from the sequence $(\varphi_\nu)$ so that

$$|\langle \varphi_1', \psi_2' \rangle| < \frac{1}{2} \quad \text{and} \quad |\langle \varphi_2', \psi_2' \rangle| > |\langle \varphi_1', \psi_2' \rangle| + 2 \tag{A.7}$$

Now, suppose we have chosen $\varphi_j'$ and $\psi_j'$ for $j = 1, \ldots, \nu - 1$. We choose $\psi_\nu'$ to be one of the $(\psi_\nu)$ sequence with index so high that

$$|\langle \varphi_j', \psi_\nu' \rangle| < \frac{1}{2^{\nu-j}}, \quad j = 0, 1 \ldots, \nu - 1. \tag{A.8}$$

Such choice is possible because the functions $\langle \varphi_j', \cdot \rangle$ are continuous on $E_2$ and the $\psi_\nu$ converge to zero in $E_2$. Taking into account (A.6) we can choose $\varphi_\nu'$ from the sequence $\varphi_\nu$ so that

$$|\langle \varphi_\nu', \psi_\nu' \rangle| \geq \sum_{j=1}^{\nu-1} |\langle \varphi_j', \psi_\nu' \rangle| + \nu. \tag{A.9}$$

Since $\|\psi_\nu\|_{k,2} \leq \frac{1}{2^\nu}$ for $k = 0, 1, \ldots, \nu$ and $E_2$ is complete therefore there exists $\psi$ in $E_2$ such that $\psi = \sum_{\nu=1}^{\infty} \psi_\nu'$. Further, for fixed $\nu$ we have

$$\langle \varphi_\nu', \psi \rangle = \sum_{j=1}^{\nu-1} \langle \varphi_\nu', \psi_j' \rangle + \langle \varphi_\nu', \psi_\nu' \rangle + \sum_{j=\nu+1}^{\infty} \langle \varphi_\nu', \psi_j' \rangle.$$

Taking into account (A.8) and (A.9) we obtain

$$|\langle \varphi_\nu', \psi \rangle| \geq |\langle \varphi_\nu', \psi_\nu' \rangle| - \left| \sum_{j=1}^{\nu-1} \langle \varphi_\nu', \psi_j' \rangle + \sum_{j=\nu+1}^{\infty} \langle \varphi_\nu', \psi_j' \rangle \right|$$

$$\geq |\langle \varphi_\nu', \psi_\nu' \rangle| - \left( \sum_{j=1}^{\nu-1} |\langle \varphi_\nu', \psi_j' \rangle| + \sum_{j=\nu+1}^{\infty} |\langle \varphi_\nu', \psi_j' \rangle| \right) \geq \nu - 1.$$

Therefore $|\langle \varphi_\nu', \psi \rangle| \to \infty$ as $\nu \to \infty$. But this contradicts the continuity of $\langle \cdot, \psi \rangle$ in $E_1$. $\qquad \square$

# SUBJECT INDEX

# NOTES AND REFERENCES TO THE LITERATURE

## Chapter 1

Lemma 1.4.1 and Theorem 1.5.1 are taken from R.A. Adams's book: Sobolev Spaces ([1]). Example 1.6.2 is taken from S.G. Krein's book: Functional Analysis ([18]). Example 1.12.1 appears in J.T. Marti's book: Introduction to Sobolev Spaces and Finite Element Solution of Elliptic Boundary Value Problems ([19]). The contents of Section 1.13 were originally produced by the authors in [16].

## Chapter 2

The contents of Sections 2.1, 2.3 and 2.4 are a modification of some of L. Hörmander's book: The Analysis of Linear Partial Differential Operators 1 ([11]). The proof of Theorem 2.2.1 appears here as in the paper: N. Meyers and J. Serrin, $H = W$ ([20]). Example 2.3.1 is originally due to V.C. Vladimirow in the book: Generalized Functions in Mathematical Physics ([25]). The idea of applying the representing theorem of linear continuous forms on Cartesian products of Banach spaces to proving representation theorems on continuous linear forms in the theory of distributions was earlier used in the books: J. Horwáth, Topological Vector Spaces and Distributions ([12]) and V.C. Vladimirow, Generalized Functions in Mathematical Physics ([25]).

## Chapter 3

Sections 3.1, 3.2, 3.3 and 3.4 are essentially an adaptation of some of L. Hörmander book: The Analysis of Linear Partial Differential Operators 1 ([11]). The idea of the sets regular at infinity is originally due to L. Schwartz: Theorie des Distributions ([24]).

## Chapter 4

The contents of Sections 4.1 and 4.2 are inspired by C.H. Wilcox's article: The Cauchy Problem for the Wave Equation with Distribution Data ([26]). The results of Sections 4.4 and 4.5 are taken from an article of J.P. Rosay: A Very Elementary Proof of the Malgrange-Ehrenpreis Theorem ([22]). The material in Section 4.6 is taken from G.B. Folland's books: Introduction to Partial Differential Equations ([6]) and Partial Differential Equations ([7]).

## Chapter 5

The spaces $\mathcal{D}'_{L^p}$ appear in L. Schwartz's book: Theorie des Distributions ([24]). The results in Sections 5.1, 5.2, 5.3, 5.4 and 5.5 concerning these spaces are presented in another way than in [24], simply. The concept of the Cauchy transformations was introduced by H. Bremermann for distributions belonging to $\mathcal{E}'$ and $\mathcal{O}'_\alpha$ in the book: Distributions, Complex Variables and Fourier transforms ([3]). The material presented in Section 5.6, 5.7 and 5.8 is taken from the article: W. Kierat and K. Skórnik, On Generalized Functions approach to the Cauchy Problem for Heat Equation ([14]).

## Chapter 6

Theorem 6.11.2 is originally due to E.J. Beltrami and M.R. Wohlers: Distributions and the Boundary Values of Analytic Functions ([2]). The results of Section 6.12 are partly presented in article: W. Kierat and U. Skórnik, On some Relations between the Cauchy and Fourier Transforms of Distributions ([15]). The material included in Sections 6.13, 6.15 and partly in 6.14 is published for the first time here. The results of Section 6.14 are taken from the article: W. Kierat and U. Sztaba, The Generalized Fourier Transforms of Slowly Increasing Functions ([17]). The contents of Section 6.16 were published for the first time in the article: W. Kierat, A Remark about the Infinitesimal Operator of the Cauchy Semigroup ([13]).

## Chapter 7

The theory of periodic distributions included in Sections 7.1 and 7.2 is originally due to L. Hörmander, The Analysis of Linear Partial Differential Operators 1 ([11]). It is presented in a more elementary form here. The contents of Sections 7.3 and 7.4 are modifications of the article: A. Cichocka and W. Kierat, An Application of the Wiener Functions to the Dirichlet Problem of the Laplace Equation ([5]).

## Appendix

The proof of Theorem A.1 is due to M.S. Brodski and it is in the book: I.M. Gel'fand and G.E. Shilov: Generalized Functions, volume I ([8]).

# BIBLIOGRAPHY

[1]   R.A. Adams, *Sobolev Spaces*, Academic Press, New York, London, 1975.

[2]   E.J. Beltrami and M.R. Wohlers, *Distributions and the Boundary Values of Analytic Functions*, Academic Press, New York, London 1966.

[3]   H. Bremermann, *Distributions, Complex Variables and Fourier Transforms*, Addison-Wesley Publishing Company INC, Reading, Massachusetts 1965.

[4]   T. Carleman, *L'integrale de Fourier et questions qui s'y rattachent*, Almqvist and Wiksell, Uppsala, 1944.

[5]   A. Cichocka and W. Kierat, *An Application of the Wiener Functions to the Dirichlet Problem of the Laplace Equation*, Integral Transforms and Special Functions, Vol. 7, No. 1–2, pp. 13-20, 1998.

[6]   G.B. Folland, *Introduction to Partial Differential Equations*, Princeton University Press and University of Tokyo Press, Princeton, New Jersey, 1976.

[7]   G.B. Folland, *Partial Differential Equations*, Springer-Verlag, Berlin, Heidelberg, New York, Tokyo, 1983.

[8]   I.M. Gel'fand and G.E. Shilov, *Generalized Functions*, Volume 1, Academic Press, New York, London, 1964.

[9]   I.M. Gel'fand and G.E. Shilov, *Generalized Functions*, Volume 2, Academic Press, New York, London, 1968.

[10]  H.J. Glaeske, *On the Wiener–Laguarre Transform of Generalized Functions, Generalized Functions and Convergence*, (Editors P. Antosik and A. Kaminski, Memorial Volume for Professor Jan Mikusinski) World Scientific Publishing Co. Pte. Ltd., Singapore, New Jersey, London, Hong Kong, 1990.

[11]  L. Hörmander, *The Analysis of Linear Partial Differential Operators 1 ( Distribution Theory and Fourier Analysis)*, Springer-Verlag, Berlin, Heidelberg, New York, Tokyo, 1983.

[12]  J. Horváth, *Topological Vector Spaces and Distributions*, Volume 1, Addison-Wesley Publishing Company INC, Reading, Massachusetts, 1966.

[13]  W. Kierat, *A Remark about the Infinitesimal Operator of the Cauchy Semigroup*, Integral Transforms and Special Functions, Vol. 4, No. 3, pp. 243–248, 1996.

[14]  W. Kierat and K. Skórnik, *On Generalized Functions approach to the Cauchy Problem for Heat Equation*, Integral Transforms and Special Functions, Vol. 2, No. 2, pp. 107–116, 1994.

[15]  W. Kierat and U. Skórnik, *On some Relations between the Cauchy and Fourier Transforms of Distributions*, Integral Transforms and Special Functions, Vol. 3, No. 4, pp. 269–278, 1995.

[16]  W. Kierat and U. Sztaba, *Differential Equations of the Second Order with Measures as Coefficients*, Annales Mathematicae Silesianae 10, Katowice, pp. 79–85, 1996.

[17]  W. Kierat and U. Sztaba, *The Generalized Fourier Transforms of Slowly Increasing Functions*, Demonstratio Mathematica, Vol.XXX, No. 2, pp. 425–428, 1997.

[18]  S.G. Krein, *Functional Analysis*, (Russian), Nauka, Moscow, 1964.

[19]  J.T. Marti, *Introduction to Sobolev Spaces and Finite Element Solution of Elliptic Boundary Value Problems*, Academic Press, London, New York, Sydney, Tokyo, Toronto, 1986.

[20]  N. Meyers and J. Serrin, $H=W$, Proc. Nat. Acad. Sci., USA, 51 pp. 1055–1056, 1964.

[21]    M. Reed and B. Simon, *Methods of Modern Mathematical Physics*, Volume 2, Academic
        Press, New York, San Francisco, London, 1975.

[22]    J.P. Rosay, *A Very Elementary Proof of the Malgrange-Ehrenpreis Theorem*, The American
        Math. Monthly, vol. 98, No. 6, pp. 518–523, 1991.

[23]    W. Rudin, *Real and Complex Analysis*, Mc Graw-Hill Book Company, New York, Toronto,
        Sydney, London, 1966.

[24]    L. Schwartz, *Theorie des Distributions*, Hermann, Paris, 1966.

[25]    V.C. Vladimirov, *Generalized Functions in Mathematical Physics*, (Russian), Nauka,
        Moscow, 1976.

[26]    C.H. Wilcox, *The Cauchy Problem for the Wave Equation with Distribution Data: on
        Elementary Approach*, The American Math. Monthly, 98, pp. 401–410, 1991.

[27]    A.H. Zemanian, *Generalized Integral Transformations*, Interscience Publishers, New York,
        1968.

# INDEX OF SYMBOLS